T0395058

Einstein's Brain

"What an intriguing achievement! In a provocative and at times chucklesome discussion across six chapters, Restivo develops the idea that 'Einstein' and 'Einstein's brain' in everyday (and even in scientific) usage are as much grammatical illusions as concrete objects of experience. Restivo encourages us to rethink the individualized sources and attributed value of being an intellectual pioneer. This book proposes a strong sociological alternative to reigning brains-in-a-vat and brain-centric ideas that define Einstein and Einstein's brain as singular and iconic achievements. Restivo doesn't diminish Einstein's uniqueness, he just situates it socially and culturally. An altogether imaginative argument and enjoyable read."

—Jaber. F. Gubrium, *Professor of Public Health, College of Nursing, University of Massachusetts and Professor Emeritus, University of Missouri, USA*

"This book is an intriguing and erudite treatise informed by sweeping knowledge of intellectual currents in multiple academic disciplines. From a decidedly sociological perspective, Restivo argues that the self is dependent on a person's social contexts, so much so that the very existence of an individual self is in question. Most of the book is not primarily about Einstein or his brain, but Restivo very cleverly uses Einstein's brain repeatedly to make his point that we ought not consider ourselves to be merely our brains. This free-ranging text represents a significant extension of the social brain hypothesis, and will be most appreciated by those with a solid background in sociology and the neurosciences, and those interested in a theoretical argument grounded in both intuition and empirical observation. Restivo's newest work is a bold re-imagining of the nature of the self, an important antidote to the long-lived but incomplete conception of genius as localized within a gifted individual's brain."

—David. S. Moore, *Professor of Psychology, Pitzer College and Claremont Graduate University, USA, and author of* The Dependent Gene: The Fallacy of "Nature vs. Nurture" *(2002); and* The Developing Genome: An Introduction to Behavioral Epigenetics *(2015)*

Sal Restivo

Einstein's Brain

Genius, Culture, and Social Networks

Sal Restivo
Ridgewood, NY, USA

ISBN 978-3-030-32917-4 ISBN 978-3-030-32918-1 (eBook)
https://doi.org/10.1007/978-3-030-32918-1

This Palgrave Pivot imprint is published by the registered company Springer Nature Switzerland AG.
The registered company address is: Gewerbestrasse 11, 6330 Cham, Switzerland

Prologue

In the early years following the publication of Albert Einstein's 1905 papers, a lot of the buzz centered around the debate about the ether in physics. The *New York Times* of September 8, 1913, reported a rift between Einstein and Max Planck on one side of the issue and Sir Oliver Lodge and his British colleagues on the other. Lodge believed in the ether, a medium eliminated in what was now known as Einstein's special relativity theory.

Ether theories have ancient roots. In the early modern and modern context a space filling transmission medium called the **ether** was believed necessary for the propagation of electromagnetic or gravitational forces. Einstein's name next appeared in the *Times* on June 10, 1918, with a story about an effort afoot to test Einstein's general relativity theory. This was a reference to predictions in Einstein's theory concerning the bending of light in gravitational fields. Evidence for the theory was already observed in the anomalous precession of the perihelion of the planet Mercury. The most dramatic and successful test of the theory came when scientists observed star light bending in the vicinity of the sun's gravitational field in 1919, an observation made possible by a solar eclipse. The paper quoted W.W. Campbell of the Lick Observatory in California on the test. Campbell was cautious about whether the test would provide meaningful data. In both stories, there was little of substance concerning the contents of the special and general theories.[1]

In the wake of the 1919 successful test of Einstein's theory, Einstein's name began to appear with some regularity in the *New York Times* and other newspapers. One recurring theme was that readers should not expect to understand Einstein's theories. Stories about Einstein's biography began to appear along with efforts to explain relativity theory to the lay public. A *Times* correspondent interviewed Einstein in December 1919. As Einstein tried to speak plainly about his theory, a clock chimed announcing the end of the interview. The clever reporter noted the irony of the absolute tyranny of time and space triumphing over Einstein's contemptuous denial of their existence.

Following an announcement by the Royal Astronomical Society endorsing Einstein's theory, an opinion piece in the *Times* compared the growing accommodation to Einstein's theory with the embrace of Bolshevism in Russia. In November 1919, Charles Poor of Columbia University wrote in that paper that there was as much conflict at the center of science as there was in the realm of society and politics. In the following months and years, the *Times* published stories on the complexity of Einstein's theories. Notables from Warren G. Harding to Einstein's wife Elsa were reported to be unable to grasp the nature and meaning of Einstein's theories.

The making of an icon and the evolution of the genius of all geniuses were well under way in the early 1920s. The *Times* continued to try to help readers understand relativity and Einstein's name started to show up in non-relativity stories. A story about French efforts in interplanetary communication included Einstein's ideas about life on other planets and how extra-terrestrial beings would communicate with us. In January 1921, Einstein's theory of the metrics of a finite universe appeared inconspicuously as a small item on the bottom of the front page of the *Times*. Now began column after column over the following years of statements by Einstein on his theories and explanatory stories by fellow physicists.

On April 19, 1955, the *New York Times* published Einstein's obituary: "Dr. Albert Einstein Dies in Sleep at 76; World Mourns Loss of Great Scientist." Thomas Harvey performed the autopsy on Einstein and without permission removed his brain, a fact unknown until 1978. It appears that Harvey's action was prompted by his knowledge of Oscar Vogt's removal and study of Lenin's brain. Einstein, concerned about possible public worshipful attention to his remains, had requested that he be cremated. He was, after Harvey removed his brain, and his ashes were scattered in and around the Institute for Advanced Study in Princeton, New Jersey.

This book is a concise introduction to the fate of Einstein's brain from a sociological perspective. I draw on research and theory at the nexus of the social-, life-, and neuro-sciences from the 1990s on that has given us a social brain paradigm. This paradigm is a radical departure from the classical view of the brain as a biological organ that could be studied independently and in isolation from social and environmental contexts—the brain-in-a-vat paradigm. In brief, the brain was classically treated as a biological organ that could be isolated in theory and research and that was the causal font of all of our thoughts and behaviors. This notion is explicitly stated as the rationale for both Bush's 1990 Decade of the Brain proclamation and Obama's 2013 BRAIN initiative. This approach was consistent with and fueled our society's reigning ways of thinking about creativity, intelligence, and genius. Given this way of thinking, it seemed reasonable to assume that the brains of geniuses who had died might hold clues to the genetic and neuronal causes of genius. The social brain paradigm in conjunction with sociological theory provides a rationale for reconsidering this idea. I take the case of Einstein's brain as the paradigmatic example of brain-in-a-vat thinking, and the failure to consider the radically social nature of human beings. I don't challenge the uniqueness of Einstein as a scientist or the uniqueness of his contributions. Rather I argue for a sociologically grounded explanation of his uniqueness, an explanation that depends on criticizing the myths of individualism, brain-centric thinking, and gene-centric thinking. I give reasons why we should replace the classic metaphor of geniuses standing on the shoulders of giants with the metaphor of geniuses (or better, as we will see, individual innovators) standing on the shoulders of social networks. My objective is to get inside of the iron cage of individualism, genius, and the brain-in-a-vat concept that is protected by the authority of a conspiracy of mythologies.

Why another book about Albert Einstein? The reason is that the large literature on Einstein trades in the currency of genius and iconic, sacred imagery. This currency overwhelms the material on Einstein as a human being. Walter Isaacson's (2008) biography on Einstein illustrates the problem I address in this book. One of many cover images you can find online emblazons the word "Genius" across Einstein's portrait in such a way as to make "Einstein" and "Genius" synonymous. The book, according to the Goodreads blurb, describes how Einstein's mind worked and what made him a genius. Isaacson finds the roots of that mind and that genius in Einstein's rebellious personality (https://www.google.com/search?client=firefox-b-1-d&q=Einstein%2BIsaacson). In the biography itself there is an interesting discussion of communications and collaborations between Einstein and his friends Michele Besso and Michael Grossman during Einstein's development of the General Theory. As in many other discussions about Einstein, we find explanations in this book that juxtapose individual personality and social relationships. Consider, alternatively, Hélène Mialet's (2010) study of the "genius" of our own era, Stephen Hawking. The subtitle of her book is

"Stephen Hawking and the Anthropology of the Knowing Subject." The contrast between Isaacson and Mialet could not be more striking. For Isaacson, Einstein's genius is rooted in personality; for Mialet, Hawking's genius is rooted in social networks, the assembly of people, ideas, and objects that made Hawking "work."

Note

1. This is a good point at which to intercede on behalf of the sociology of science and to alert the reader to the complexity of truth, objectivity, proof, and experiment. Two sociologists of science, Harry Collins and Trevor Pinch (1998), studied the history of the solar eclipse experiment and reported their findings in an article titled "Two Experiments That 'Proved' the Theory of Relativity." Why did they put scare quotes around the word "proved"? The reason is that empirical sociological studies of science and scientific knowledge from the late 1960s on demonstrated that the process of science was fallible, untidy, corrigible, and as much art and craft as rationality and logic. The point was not to undermine the facticity of the findings of science but to show that the facts and knowledge produced by science were not products of a pure, unadulterated, logic; the very ability of scientists to produce true and useful knowledge was in fact just a messy process by its very nature. Collins and Pinch (1998: 54) conclude their study of the Michelson and Morley ether experiment and the solar eclipse observational study led by Sir Arthur Eddington as follows:

   > We have no reason to think that relativity is anything but the truth—and a very beautiful, delightful and astonishing truth it is—but it is a truth which came into being as a result of decisions about how we should live our scientific lives, and how we should license our scientific observations; it was a truth brought about by agreement to agree about new things. It was not a truth forced on us by the inexorable logic of a set of crucial experiments.

   As you read through the science of Einstein and his brain, Collins' and Pinch's conclusion should be kept in mind. Observation, experiment, proof, and replication are not misleading us in science; but they are far more complicated processes than the lay public and even many scientists themselves realize. Science is not and cannot be practiced by one person, or one organization; you can't do science in one laboratory. Science cannot establish facts in a single case, observation, proof, or experiment. Science works its way to different levels of facticity, closure, and truth collaboratively over time and across generations.

Acknowledgments

I have acknowledged the shoulders of the social networks I stand on in detail in my earlier publications. Here I want to express my special indebtedness regarding my work on brain and mind to Wenda Bauchspies, Colin Beech, Leslie Brothers ("Sal Restivo is still the King of Cool"), Mary Ann Castle, Randall Collins, Jennifer Croissant, Peter and Daniel Denton, Rachel Dowty, Ellen Esrock, Julia Loughlin, Hélène Mialet, Ann Miller, Lew Pyenson, Kaia Raine (née Karl Francis), Hilary and Steven Rose, Elizabeth Parthenia Shea and her "Groove," Alex Stingl, and Sabrina Weiss. Sal Restivo.

Jenelle Clarke wishes to acknowledge and thank the therapeutic communities involved in her research study. Thanks also to Professor Nick Manning, Dr. Gary Winship, and Professor Ruth McDonald. This work was supported by the Economic and Social Research Council (grant number ES/J500100/1]) and the Institute of Mental Health (Nottingham).

Contents

List of Figures

CHAPTER 1

"Einstein" as a Grammatical Illusion

Abstract This chapter introduces the rationale for challenging the view of Einstein as an icon and genius. We, along with Einstein and others, have made a mistake in reference. The tendency has been to view him as a unique individual and to look to genes and neurons (and more broadly biology) to explain his uniqueness. The label "genius" adds a divine factor to the explanatory narrative. I do not suggest that we deny his uniqueness but that we situate it socially, culturally, and historically. From the vantage point of social science, "Einstein" is defined by the particular configuration of social networks he engaged as his life unfolded. The chapter also previews the following chapters on brains, self, and genius.

Keywords Brain • Genius • Culture • Network • Self

Introduction

Why another book about Albert Einstein? The reason is that we have mistaken Einstein as icon and genius for Einstein the human—and more significantly—*social* being. If this were a murder mystery, I would now be giving away the identity of the murderer. We—along with Einstein himself—have made a mistake in reference. The tendency has been to view him as a unique individual and to look to genes and neurons (and more broadly biology) to explain his uniqueness. The label "genius" adds a divine factor to the explanatory narrative. I am not suggesting that we

S. Restivo, *Einstein's Brain*,
https://doi.org/10.1007/978-3-030-32918-1_1

deny his uniqueness but that we revise our understanding of that uniqueness, that we contextualize it, and that we ground it. From the vantage point of social science, "Einstein" is defined by the particular configuration of social networks that he engaged as his life unfolded.

History is awash with errors and corrections concerning mind, brains, consciousness, and self. We have made a mistake in looking for the roots of genius inside the brains of the dead and indeed have made a mistake in viewing the individual as the locus of genius. More generally, we have made the mistake of looking for consciousness and the roots of all of our behaviors and thoughts in the isolated biological brain. The very term "genius" has contributed to our mistakes. Brains living and dead and individuals do have stories to tell but their stories are not about genetic, neuronal, or divine roots of thoughts and actions. Instead, they are about lived experiences. One can already notice contradictions between those who claim that Einstein was destined to be a thinker and those who point out the influence of Einstein's father and uncles on the direction of his intellectual life. To correct the myth of individualism we have to learn to recognize the reality and power of the social and cultural forces that shape our lives.

Social scientists who argue against the myth of individualism vary in the extent to which they assign agency or free will to individuals. In part this is a function of the extent to which they take their personal experiences of the "sense of self" as introspectively transparent. Certainly, experience is in the end the final arbiter of what is going on in and around us. However, it is not individual experience that we rely on in science but the collective intersubjectively tested experience of a community over time. Historically, we have relied primarily on the generational continuity of communities of specialized knowledge makers from philosophers and natural philosophers to scientists. There are also locally and narrowly circumscribed communities of knowledge makers (e.g., women, traditional healers, bodybuilders, and activists in the LGBTQ communities). They are less likely to generate globally reliable and valid knowledge, but sometimes the knowledge they generate does get absorbed by the scientific community.

Consider that as individuals we do not experience the earth in motion. And yet it spins on its axis west to east (wobbling in precession) making one full turn roughly every 24 hours; it is rotating at any point along the equator at about 1000 miles per hour; it travels around the sun at 66,000 miles an hour; it is part of a solar system orbiting the center of the Milky Way at 140 miles a second. The Milky Way is part of a cluster of galaxies (The Local Group) and is traveling toward the center of the cluster at 25 miles a second. And The Local Group itself is speeding through space at 370 miles a second. None of this motion is accessible to individual experience. And yet we have knowledge of these motions through the collective generationally linked intersubjectively tested experiences of scientists. Since my approach and perspective are grounded in the norms and practices of science, it is important to point out the fragility and vulnerability of science. Intersubjective testing is a social process and subject to social forces that can interfere with its function in grounding objectivity. For example, bureaucratization and professionalization can interfere with this function. Therefore it is important not to take science, intersubjectivity, or the supposed self-correcting nature of science for granted (Restivo 1975).

In order to account for the introspective experience of agency, some social scientists have argued that there is some sort of "slippage" between "individual" and "society" where agency lives. I am more persuaded by social structural patterns of behavior and thought than I am by the illusion or fallacy of introspective transparency. I am at the radically structural end of the continuum on this matter, granting individuals no causal or willful control over their own lives. I do not challenge the experience of agency and free will but I attribute this to social and cultural configurations and not to the reality of these aspects of the sense of self. We can account for the experience by drawing on the concept of complex open systems which are variously indeterminate but lawful.

Einstein and his brain are iconic examples of the myth that we are our brains and that we are free-willing individuals. That is why they are the focus of this book. To challenge the myth that we are our brains and that we are individuals in the strong sense, we have to engage the social brain paradigm and the evolution of humans as the most social of the social species. This is the theme of Chap. 2.

Guide Posts on the Road to a Sociology of Einstein and His Brain

The Politics of Einstein

The politics of Einstein as an icon of science is about linking the social relations of Einstein with the social relations of science. What should we make of all those photos of him with leading political leaders from all over the world, and with leaders of the World Zionist Organization? Is there something sinister revealed about the relationship between science and society in all those photos of him posing with political leaders from Ben-Gurion to Ramsey Macdonald, from German political and business leaders to Churchill and Lloyd George, from Alfred E. Smith to Queen Elizabeth of Belgium, President Harding to King Albert, and from Harry Truman to Einstein's correspondent Ataturk? There is no need to impugn Einstein's motives or the humanistic spirit that is behind some or all of these photos. But the humanistic side of these photos is a phenomenon of the individual and in particular and without doubt of Einstein himself. The photos, however, also reveal the nature of science as a tool of the ruling classes. The fact that in many of these photos we can feel the shadow of Adolf Hitler should make us wonder if these political figures represent the benevolence of the state and its love of science or as their understanding of science as a force for national defense and war. In recent times this relationship has become more complicated most notably in the United States as anti-science attitudes and agendas have nested in all parts of our society and penetrated all the way to the Oval Office.

I Meet Einstein's Brain, I Meet Einstein

In order to understand the myth of the "I" we need to explore the notion of the social self. To say that the "I" is a myth and an illusion is to say that the very idea of a free-standing free-willing individual is incompatible with our radically social nature. To introduce the rationale for labeling the "I" a grammatical illusion, let's begin in an Einsteinian way with an exercise of the imagination. Imagine walking into the Museo Galileo (formerly the Instituto e Museo di Storia della Scienza) in Florence, Italy, coming across Galileo's middle finger in a display case, and remarking "I meet Galileo's middle finger, I meet Galileo." Now imagine viewing what is supposed to be Napoleon's penis inside a box in a private home in New Jersey and remarking "I meet Napoleon's penis, I meet Napoleon." These are imagined encounters. But the Japanese mathematician Kenji Sugimoto believed that when he "met" Einstein's brain (at the end of the day for him, a small piece of tissue cut off from the remains of Einstein's brain), he met Albert Einstein. Unlike the first two examples, this example is real. All three cases are examples of relicism, the idea that encountering portions of a deceased person is the same as encountering that person in some way and in the extreme actually equivalent to meeting that person. This seems like utter nonsense in the case of Galileo's finger and Napoleon's penis but somehow more plausible in the case of Einstein's brain.

When Dr. Thomas Harvey took Einstein's brain (without permission) during his autopsy on the famed scientist's corpse, he set in motion decades of efforts to find in the dead tissue of Einstein's brain the secret to his genius. Einstein's brain moved into the landscape of urban legends. His brain actually was removed by Dr. Thomas Harvey during the autopsy. But Harvey did not disappear, and Einstein's brain did not end up in a garage in Saskatchewan. The story of Einstein's brain is told in many different works. One of the most interesting and accessible accounts is Michael Paterniti's *Driving Mr. Albert: A Trip Across America with Einstein's Brain* (2000). Harvey took the first step in relicizing Einstein's brain.

Hundreds of hours have been spent during the decades since Einstein's death in 1955 examining the dead tissue of his brain with the objective of identifying the neuronal origins of Einstein's genius. During this period, we have made tremendous advances in our knowledge of the brain. As we will see, these advances have tended increasingly to demonstrate that the brain is not a free-standing autonomous causal source of our behaviors and thoughts. Even though this idea continues its strong hold on neuroscience and the popular imagination, the evidence for environmental and social influences on the brain is one of the hallmarks of frontier research in neuroscience and in particular neuroplasticity since the 1990s. It should be obvious that the brain is a better place to look for the roots of genius than the middle finger or the penis. But is the brain in fact where genius resides, alongside God, morals, and sexual preference? And do we really know what we're looking for when we go looking for genius inside or outside of the brain? My objectives in this book are in part to show why we've been looking in the wrong place for genius and why the very idea of genius itself should give us pause.

I am not going to unfold a detailed biography of Albert Einstein. His life and times have been written about for decades and by such outstanding biographers as Ronald Clark and Walter Isaacson. It is true that these biographies promote to different degrees distorted and erroneous images about Einstein. I bracket the facts of Einstein's life, which the better biographers are more often than not right about. I want to focus here on how Einstein has been described, on the concept of Einstein's self that comes through in his biographies. How do these views of Einstein match up with what we know about the self as a sociological idea? I discuss the theory of the social self in Chap. 2.

Let's look at some typical descriptions of Einstein in the literature. Einstein is described as a 26-year-old "wonder" who produced "astonishing" theories, a scientific supernova, the embodiment of genius composed of character, imagination, and creativity, someone born to and indeed destined to become a thinker, and the genius of all geniuses. What has elevated him to this status? For one thing, he belongs to our world; we can see and hear him, we can read about the accolades, the fact that he has changed the most fundamental ways in which we understand space and time. God seems to look out at us through his soulful eyes. He escapes all bounds of convention with his baggy clothes and shock of white hair going off in all directions.

A sober evaluation would have to check his right to the title of "genius of all geniuses" against those of Newton and Leibniz. Newton transcended all earlier versions of what it meant to be a scientist (using a word that was not yet invented in his time); the calculus may have been the least of his contributions. And Leibniz, "the continental Newton," has been described by his biographers as "the last universal genius" and "the most comprehensive thinker since Aristotle". What do the biographical descriptions of Einstein actually tell us about Albert Einstein the man? What is the mythology that drives such descriptions? Einstein himself puts that mythology into words when he claims that new ideas can only be produced by individuals.

We are obliged to acknowledge without diminishing in any way Einstein's achievements that these descriptions are grounded in the myth and cult of individualism. I take the title of this chapter from various remarks by Nietzsche that can be summarized in the idea that the "I" is a grammatical illusion. The illusion is in a grammar that identifies each one of us in terms of the first person singular subject or the person referred to as the grammatical subject of a sentence. In order to understand Einstein as a person as opposed to an icon, myth, or most importantly a singular subject, we need to understand persons as social entities. Only then can we see the mistake we—along with Einstein himself—have made in trying to understand Einstein the man. The tendency has been to view him as a genetically and neuronally unique individual who was divinely inspired. I am not suggesting that we deny his uniqueness but that we revise our understanding of that uniqueness sociologically. We can pursue this line of inquiry further by exploring the very idea of "genius." I will explore this in detail in Chap. 4.

Genius and Culture

When we identify Einstein as a genius or the genius of all geniuses, we learn more about our culture and ourselves than we do about Einstein. First of all, the term "genius" enhances and exaggerates the concept of the individual as an entity that stands apart from society, history, and culture, and even from time and space. The element of the divine spins the genius right out of the world itself into some transcendental and sacred space. Second, it sets Einstein apart from the rest of us in an intellectual world we could never hope to understand let alone inhabit ourselves. For the concept of "genius" to be meaningful, for it to mean scientifically what it conveys to the general public in everyday terms, it would have to be rooted in genes, neurons, or both. In that case, geniuses would appear at random and scattered across the intellectual landscape. On the contrary, the most comprehensive studies of genius by social scientists have demonstrated that even if we accept the validity of the concept of genius geniuses do not appear at random. Genius clusters. And it clusters at particular civilizational and cultural historical junctures. I discuss this further in Chap. 4.

However problematic the term itself is, the application of "genius" has a cultural context. A good example of this is "the" calculus. Many if not all mathematics textbooks refer to Newton and Leibniz as the co-inventors of "the" calculus. In fact, they invented (some would say discovered) two different calculuses. One distinguished historian of science described their methods as "profoundly different." The technical details go beyond what I can consider here. But there is no disputing Newton's priority in inventing a new infinitesimal calculus and developing it into differential and integral calculus. Newton's approach was geometrical, inductive, and intuitive, whereas Leibniz's drew on the more formal deductive methods of algebra and logic. Furthermore, sociologists of science have demonstrated that they approached their subject matter in ways that reflected their positions in the religious, social, and political structures of their time.

The difference between the two calculuses is at bottom conceptual. They lead to the same results but the conceptual differences are not to be ignored. Newton apparently used his calculus to arrive at his most important results and then reworked them so they could be presented in terms of traditional geometrical (Archimedean) methods. Newton's success was not simply a matter of his intellect but a function of his power and influence at the court of Prince George and Queen Ann. Newton's interests converged with the practical problems of his time involving the dynamics of change in navigation and commerce. Leibniz had more in common with Einstein in his quest for universal unifications in mathematics, science, and language. His priority dispute with Newton, like priority disputes in general, was sustained by champions and acolytes. Priority disputes are a little strange because they assume the myth of individualism and the independence of genius. Who gets credit in the end is more often nothing more complicated than being in the right place at the right time and having the strongest network of supporters. There is finally one more interesting point to be made here. Randall Collins and I have shown that major priority disputes, conflicts, and scandals in mathematics signal shifts in the organizational and institutional structure of mathematics. In the case of Newton and Leibniz, the shift is from traditional forms of patronage to government sponsorship and the emergence of scientific academies. The case of Newton and Leibniz and others like it help explain some of the criticisms of the uniqueness of Einstein's discoveries and charges of plagiarism. This relationship between major scandals and organizational shifts may be a general feature of the sciences.

The story I grew up reading as a "science nerd" was that Einstein, isolated from the major centers of activity in physics as a clerk in a Swiss patent office, suddenly burst upon the world scene with his theory of relativity in 1905, shocking the worlds of physics, scientists and intellectuals, and the lay public. As an undergraduate and later in graduate school, my readings in the history of science introduced me to the works of the central figures in late nineteenth- and early-twentieth-century physics and mathematics. I learned that relativity theory had been a part of the history of physics long before Einstein came along. Indeed, if we define relativity theory broadly enough, its history goes back to the origins of the scientific revolution and beyond. There were so many precursors to Einstein's work that the charge of plagiarism emerged in the minds of some physics watchers. Most of these charges can be tied to the Nazi sympathizer and anti-Semite Phillip Lenard. As an advocate of "German physics," he opposed Einstein's "Jewish physics."

There is something infantilizing and anti-intellectual about the way we are schooled on important figures in our history. We are taught icons not human beings. We are taught heroes and heroines and geniuses because this serves to imbed us in a national spirit of individual achievement driven by the myth and the cult of individualism. If our education stops at the end of the normal high school term, we will likely carry those schooled views of sanitized individual achievements into our adult lives. It is only by pursuing higher education and especially by developing a life driven by inquiry (and not necessarily formal schooling) that we discover human beings behind the icons, myths, and sometimes lies driving our childish thoughts about Santa Claus, the tooth fairy, George Washington, and the gods.

I have told my students, many of whom believe in genius and the cult of the individual, that if they give me a genius I will show them a network. What is the Einstein network?

The Einstein Network

In a very general sense, the forms of relativity theory can be traced back to Da Vinci, Galileo, and Giordano Bruno. In Einstein's time, Poincaré discussed the principle of relativity in papers presented in 1904 and 1905. In these papers, Poincaré had energy-mass equivalence, an equivalence already published by Olinto De Pretto in a 1903 paper. Samuel Taylor Preston had speculated on mass-energy equivalence as early as 1875. We have in these examples the core of the multiples context within which Einstein worked. Inventions and discoveries tend to emerge in different places at the same time as families of simultaneous innovations (see Chap. 3). The network that is Einstein has different levels. There is the primary level of his close friends and associates including notably his wife Maria, and his friends Besso and Grossman. The secondary level includes those scientists he met with face to face such as his teachers and mentors and those he corresponded with; and there is a tertiary level of scientists, dead and alive, whose works he read.

Einstein's 1905 papers come in the midst of a cultural flowering of ideas, inventions, and discoveries across the full spectrum of the arts, humanities, and sciences between 1840 and 1920. Einstein's "genius" cluster in physics included such luminaries as Planck, Tesla, Marconi, Westinghouse, Madame Curie, the Wright Brothers, and Edison. The two great innovations in physics that would remain at the core of physics throughout the twentieth and into the twenty-first century, relativity theory and quantum mechanics, are born in the early years of the twentieth century. Expanding the genius cluster to encompass music and focusing on the turn of the century, we can include such names as Sibelius, Puccini, Debussy (who introduced impressionism in his *Pelleas and Melisande* in 1902), Schoenberg (dissonance; his first 12 tone work was Piano Suite, Opus 25, 1921), and Stravinsky (*Firebird*, 1910; *Rites of Spring*, 1913), and Charles Ives (Fourth Symphony, 1916).

In literature, we have the Victorian period (1832–1901) and the rise of the novel; American Transcendentalism (1836–1860; Emerson), Realism (1865–1914; Flaubert, Tolstoy), Stream of Consciousness (early twentieth century; Joyce published Ulysses in 1918; Woolf); various forms of Modernism (early twentieth century), Naturalism (1900–1930, Zola), Edwardianism (1901–1910, division between high and low literature and the growth of children's literature), and the Harlem Renaissance of the 1920s (Hughes and Hurston). D.H. Lawrence published *Sons and Lovers* in 1913; Butler's *The Way of All Flesh* appeared in 1913; Henry James was active in the first decade of the twentieth century; Dreiser's *Sister Carrie* appeared in 1900; and Conrad, Wharton, Maugham, Forster, and London were prominent literary voices in the early 1900s. Tarkington's *The Magnificent Ambersons* was published in 1918. And Eliot's *Prufrock* appeared in 1915, to be followed in the 1920s by *The Waste Land* and *The Hollow Men*. In philosophy, we have Wittgenstein's *Tractatus Logico-Philosophicus* from 1918. There was a sympathetic mutuality that linked Cubism (see Picasso, and especially his *Les Demoiselles d'Avignon*, 1907) and Relativity Theory. They both involved challenges to conventions regarding absolute time and space. Stein associated her writing with Cubism (Perelman 1994: 135) and thus one might say with a literary transform of the theory of relativity.

The period 1840–1920 saw the emergence of what I have called The Age of the Social, a veritable Copernican revolution that witnessed the emergence and crystallization of the social sciences, broadly conceived to encompass sociology, anthropology, and social psychology. The period is marked by the names of Madame de Stael (if we stretch our period backward to include this significant contributor to European Romanticism and precursor to Marx: d. 1817), Harriet Martineau, Charlotte Perkins Gilman, Jane Addams, Saint-Simon and Auguste Comte, Marx, Weber, Durkheim, Nietzsche, and Kropotkin, among others. If we are going to play the dangerous game of conferring the term genius, then surely Durkheim, Marx, and Nietzsche must stand with the Newtons, Einsteins, and Darwins. Darwin, of course, heads the list of fashioners of a new evolutionary biology of the human species. This period also saw the development of a new and revealing historiography of Christianity alongside Durkheim's sociology of God and religion and Weber's comparative sociology of the religions of India, China, and the Jews. Durkheim gave us a striking theory of God as a symbol of society; Weber linked Calvinism and capitalism. A more detailed history of the Jews and the emergence of the

Christian era developed as new data came to the attention of European scholars and Enlightenment views spread among them.

Genius: A Brief Preview

The term "genius" is loaded with dangers and it might be wise to eliminate it from our vocabulary. The most immediate danger of course is that it fuels the myth of individualism. Etymologically, it is associated with a spirit present at birth, innate ability, and divine inspiration. The latter idea comes from the original meaning of the term, "tutelary or moral spirit attendant on a person." The genius is guided from birth on by a guardian deity. The term has been the subject of much debate and the debate has tended toward resolution by reducing the term to more readily operationalizable ideas like creativity and eminent achievement.

It is no accident that the term "genius" achieves increasing prominence from the 1500s onward coincidentally with the emergence of the capitalist model of an economy based on private property and individual (to the point of atomistic) behavior in a mythically labeled "free market." Its origins go back, of course, to ancient Rome. "Genius" allows us to erase or otherwise ignore the genius cluster and the multiples so that an invention or discovery can be assigned to a particular individual. To recall the implication of the concept of the social self, individuality is not a matter of what is inherent, genetic, or neuronal but rather of the particular social configuration that represents the social groups and networks that mark the self's progress through the life span.

To put it in the most radical (read "realistic") terms, following the Polish sociologist Ludwig Gumplowicz (1838–1909), we can say that the individual is a voice box for those groups and networks she or he has engaged over her or his lifetime. From the perspective of the sociology of knowledge, we can say that ideas are generated by and in social networks and manifest themselves in the cognitive experiences of individuals who can think thoughts and collectively construct technologies that manifest those thoughts. Finally, looking to Galton's (1822–1911) work on hereditary genius, we should not ignore the connection of the genius literature from Galton on with the eugenics movement. There is thus an element of racism in the concept of genius (as in the concept of the superior being, God, man, and so on).

The myth of genius lies primarily in its connection to the myth of individualism. I have noted that the myth of individualism is fueled by and fuels the ethos of capitalism. The relationship between the coincident emergence of the capitalist ethos and the modern concept of genius has been noted by others. So has the relationship of the modern concept of genius to the emergence of new ideas about aesthetics and the self. Darrin M. McMahon, in his 2013 book *Divine Fury*, argued that the original ties of the concept to divinity continue to be the primary fuel for our fascination with genius. All of these factors are part of the fabric that gives genius its power over our imagination and all are grounded in the simultaneous integrated emergence of the capitalist ethos, Protestantism, and modern science.

There is no question that in its everyday command of our attention genius does play on our awe of, obeisance to, and worship of the genius as someone who carries something of the divine and sacred within himself or herself. Notice that it has been necessary to make a specific case for attributing genius to women; recall how radical it was for Gertrude Stein to claim "I am a genius":

> As a young woman, Stein became interested in the writings of the Austrian philosopher Otto Weininger, whose work *Sex and Character* (1903) "argued that both 'the Jew' and 'the woman' were the negation of the ideal and universal type of 'genius.'" He identified men of the "Germanic races" as those most capable of exhibiting genius. (Doyle 2018: 44)

Using the form of the domestic memoir (*The Autobiography of Alice B. Toklas*), Stein merged the identities of "wife" and "genius" making genius collaborative and domestic, and challenging the association of genius with male autonomy. At the same time, she provides some support for the concept of the "I" as a grammatical illusion. There is, of course, a patriarchal thread that holds the tapestry of capitalism, Protestantism, and science together and thus contaminates the very idea of genius. I will discuss this further in Chap. 4 under the heading "Gaging Gender and Genius."

Conclusion

Let's return now to the question I opened this chapter with: Why another book on Einstein? While I was developing the idea for this book, *National Geographic* published a special issue on "genius" in May 2017. This issue appeared in conjunction with a ten-part television series on their cable channel. The first paragraph of the opening article by Claudia Kalb almost immediately turns to Einstein:

> With no tools at his disposal other than the force of his own thoughts, he predicted in his general theory of relativity that massive accelerating objects—like black holes orbiting each other—would create ripples in the fabric of space-time.

A century later, using "enormous computational power, and massively sophisticated technology," gravitational waves were detected confirming his theory. Keep in mind that a minority of scientists is skeptical of this discovery; and we cannot be sure that what Einstein's mathematics and vision revealed were what we now know as black holes and gravity waves. The operative words here oppose the force of Einstein's "own" thoughts to the "enormous" and "massive" power of modern computers and technologies. Herein lies the provocation for the question behind the special issue: "What makes a genius?" Where in this idea of Einstein's "own" thoughts is there room for his intimate association with Besso and especially Grossman who helped Einstein with the non-Euclidean (mainly those associated with Riemann) ideas and the concept of tensors he needed to formulate the general theory?

The modern idea of genius emerges in the context of and reflects the ethos of capitalism. The emergence of new views about aesthetics and the self has also been pointed to as giving us the modern view of genius. McMahon argues persuasively that the original divine roots of genius continue to be the major fuel energizing the way we view the genius. The modern genius, like the ancient one, has something of the divine and the sacred about him or her. We saw earlier that a special case has had to be made for applying the term "genius" to women; Gertrude Stein was provoked to announce her genius out loud. The male color of genius is readily explained by the fact that the thread that holds the tapestry of science, economy, and religion together is patriarchy. McMahon has certainly hit on something in terms of the everyday person's experience of awe, idolatry, and worship on encountering the genius; it is like encountering a god when you meet Edison or Einstein. This is however part of the general political economy that undergirds the myth of individualism and supports the social role of the genius. McMahon argues that the genius is disappearing from the cultural landscape and that Einstein might be the last genius. Perhaps, playing off the divinity of genius, if God is dead, genius is if not dead moribund.

My objective in this chapter has been to plant some seeds of discontent with the very idea of genius because it is problematic and should not be taken for granted. This has been noted by many observers of the genius myth. I also want the reader to be uncomfortable with the idea of individualism. The path past this myth is the concept of the social self and that is the topic of the next chapter.

Bibliographic Notes for Chapter 1

The "I" as a grammatical illusion

For a discussion, review, and relevant citations regarding Nietzsche's views on the "I" as a grammatical illusion, see Gardner (2009).

Biographies of Einstein and Einstein's Brain

For biographies of Einstein, I recommend Isaacson (2008) and Clark (1971/1984). For a journalistic biography of Einstein's brain, see Paterniti (2000). For a neurological story of Einstein's brain, see Lepore (2018).

Relics

On a book about the relics of the Catholic Church that has the stamps of the Nihil Obstat and Imprimatur, see Cruz (1984); Bagnoli, Klein, Mann, and Robinson (2010) is an overview of relics and culture; on the relics of science, see Beretta, Confori, Mazzarello (2016).

Newton, Leibniz, and "the" Calculus

Describing what Newton and Leibniz had in common, Cambridge's Simon Schaffer began by saying they both have biscuits named after them. I don't know which one invented calculus, but I do know which biscuit I would rather have with my afternoon coffee! Johnston (2009). On the history of the Newton-Leibniz controversy, see Guicciardini (1999); Sonar (2018); Shapin (1981); Westfall (1980); Collins and Restivo (1983, esp. pp. 205–212).On the political economy of Newton's *Principia,* see Hessen (1931). That this is a fundamentally sound sociological treatise and not some piece of vulgar Marxist rhetoric and propaganda is demonstrated in Merton (1968). Merton clarifies the distinction between the personal motivations of scientists (the basis for historian G.N. Clark's criticism of Hessen) and the structural determinants of their research. A succinct literate demonstration of what it means to think sociologically.

Energy-Mass Equivalence and the Speed of Light: The Paths to Einstein

- Web pages:
  Koberlein (2017); Ball (2011); Rothman (2015).
- Books and Articles:
  Corry, Renn, and Stachel (1997); Winterberg (2006; response to Cory-Renn-Stachel); Bodanis (2009); Cox and Forshaw (2009); Rathore (2018).

The Einstein Network and the 1840–1920 Genius Cluster

The material on the Einstein network and the 1840–1920 genius cluster is based on independent research using a variety of website and historical overviews. See the references for Energy-Mass Equivalence and the Speed of Light: Before Einstein; and see Whittaker (1989); Miller (2001); and Restivo (2018).

Supplementary References

On the sociology of the "individual," see Gumplowicz (1980); on the very idea of "genius," see McMahon (2013); and for the key article in the National Geographic special issue on "Genius," see Kalb (2017).

References

Bagnoli, M., H.A. Klein, C.G. Mann, and J. Robinson, eds. 2010. *Treasures of Heaven: Saints, Relics, and Devotion in Medieval Europe*. New Haven: Yale University Press.

Ball, P. 2011. Did Einstein Discover E=mc^2? (/E=mc^2/HistoryOfE=mC2A.webarchive).

Beretta, M., M. Confori, and P. Mazzarello, eds. 2016. *Savant Relics: Brains and Remains of Scientists*. Sagamore Beach, MA: Science History Publications.

Bodanis, D. 2009. *$E = mc^2$: A Biography of the World's Most Famous Equation*. New York: Bloomsbury.

Clark, R. 1971/1984. *Einstein: The Life and Times*. New York: Avon Books.

Collins, R., and S. Restivo. 1983. Robber Barons and Politicians in Mathematics: A Conflict Model of Science. *The Canadian Journal of Sociology* 8 (2): 199–227.

Corry, L., Jürgen Renn, and John Stachel. 1997. Belated Decision in the Hilbert-Einstein Priority Dispute. *Science* 278 (5341): 1270–1273.

Cox, B., and J. Forshaw. 2009. *Why Does E=mc^2?* Boston: Da Capo Press.

Cruz, J.C. 1984. *Relics*. Huntington, IN: Our Sunday Visitor.

Doyle, N. 2018. Gertrude Stein and the Domestication of Genius in The Autobiography of Alice B. Toklas. *Feminist Studies* 44 (1): 43–69.

Gardner, S. 2009. Nietzsche, the Self, and the Disunity of Philosophical Reason. In *Nietzsche on Freedom and Autonomy*, ed. Ken Gemes and Simon May, 1–31. Oxford: Oxford University Press.

Guicciardini, N. 1999. *Reading the Principia: The Debate on Newton's Mathematical Methods for Natural Philosophy from 1687–1736*. Cambridge: Cambridge University Press.

Gumplowicz, L. 1980. *The Outlines of Sociology*. New Brunswick, NJ: Transaction Books, 1980; orig. publ. in German, 1885.

Hessen, B. 1931. The Social and Economic Roots of Newton's Principia. In *Science at the Crossroads*, ed. N.I. Bukharin et al., 151–212. London: Frank Cass and Company.

Isaacson, W. 2008. *Einstein: His Life and Universe*. New York: Simon & Schuster.

Johnston, H. 2009. Newton's Wars. *Everyday Science/Blog*, September. https://physicsworld.com/a/newtons-wars-ii/.

Kalb, C. 2017. What Makes a Genius? *National Geographic* 231 (5): 30–55.

Koberlein, B. 2017. The History of Einstein's Most Famous Equation. https://briankoberlein.com.

Lepore, F.E. 2018. *Finding Einstein's Brain*. New Brunswick, NJ: Rutgers University Press.

McMahon, D.M. 2013. *Divine Fury: A History of Genius*. New York: Basic Books.

Miller, A.I. 2001. *Einstein, Picasso: Space, Time, and the Beauty That Causes Havoc*. New York: Basic Books.

Paterniti, M. 2000. *Driving Mr. Albert: A Trip Across America with Einstein's Brain*. New York: Dial Press.

Perelman, B. 1994. *The Trouble With Genius: Reading Pound, Joyce, Stein, and Zukovsky*. Berkeley: University of California Press.

Rathore, H. 2018. Why Did Einstein Use Speed of Light Squared in the Famous Equation $E = mc^2$? https://www.quora.com/Why-did-Einstein-use-speed-of-light-squared-in-the-famous-equation-E-mc-2.

Restivo, Sal. 1975. Towards a Sociology of Objectivity. *Sociological Analysis and Theory* 5 (2): 155–183.

Restivo, S. 2018. *The Age of the Social: The Discovery of Society and the Ascendance of a New Episteme*. New York: Routledge.

Rothman, T. 2015. Was Einstein the First to Invent $E = mc^2$?(/E=mc2/Was%20Einstein%20the%20First%20to%20Invent%20E%20=%20mc2%3F%20-%20Scientific%20American.webarchive).

Shapin, S. 1981. Licking Leibniz. *History of Science* 19: 293–305.

Sonar, T. 2018. *The History of the Priority Dispute Between Newton and Leibniz: Mathematics in History and Culture*. Cham, Switzerland: Birkhäuser, orig. in German, 2016.

Westfall, R.S. 1980. Newton's Marvelous Years of Discovery and Their Aftermath. *ISIS* 71: 101–121.
Whittaker, E. 1989. *The History of the Theories of Aether & Electricity*, Vols. 1 & 2. Mineola, NY: Dover Publications; orig. publ. 1951–1953.
Winterberg, F. 2006. Response to Cory-Renn-Stachel. *Zeitschrift für Naturforschung* 59a: 715–719.

CHAPTER 2

The Social Self: Beyond the Myth of Individualism

Abstract In this chapter, I fill in the sociological theory of the self that grounds the concept of the "I" as a grammatical illusion. I introduce the idea of humans as always, everywhere, and already social. We are the most radically social of what biologists refer to as the eusocial species, the highest level of animal social organization. The chapter deals with how society constructs individuals as social matrices, the recurrence theorem (which explains our ability to sustain our self-concepts as we move across different social settings), the problems of programming, free will, and consciousness, the evolutionary context of the social self, and the limits of the Golden Rule. I also introduce the concept of dissocism, the inability to "see" the social as a nexus of causal forces (thus making "the social" invisible to large portions of populations across time, space, and culture).

Keywords Self • Society • Evolution • Individualism • Programming • Golden Rule

Introduction

In Chap. 1 I introduced the concept of the "I" as a grammatical illusion and discussed how the myth of individualism conditions our idea of Einstein as a genius. In this chapter I fill in the sociological theory of the self that is the foundation of my conception of Einstein as a grammatical

S. Restivo, *Einstein's Brain*,
https://doi.org/10.1007/978-3-030-32918-1_2

illusion. What are we to make of Einstein as one of us, a member of a radically social species?

As members of a species we are born under a social imperative—humans are always, everywhere, and already social. We are the most social of the eusocial species. Biologists use the term "eusocial" to refer to the highest level of animal social organization. This is defined according to the following characteristics: cooperative brood care (including care of offspring from other individuals), overlapping generations within a colony of adults, and a division of labor into reproductive and non-reproductive groups. Eusocial species are divided into groups sometimes called castes. Eusociality is different from all other systems which can be described as social because individuals of at least one caste usually lose the ability to perform at least one behavior characteristic of individuals in another caste. Ants, wasps, and bees are among the most prominent of the eusocial species studied by biologists. Humans are arguably considered one of the eusocial species and are the most radically social of these species.

The fact that we are the most social of the eusocial species is a big evolutionary claim. It means that we come into existence as a species already social, not as individuals who become social. How is this manifested locally in our own lives? First, we are born into cultures already awash in traditions, norms, values, beliefs, and perhaps most importantly languages. We do not get to choose our birth place, our parents, family, national origin, culture, or our language. It makes a difference in your life chances and life choices if you are born in Belgium rather than Outer Mongolia; and it makes a difference if you are born in the Flemish or the French-speaking section of Belgium. The differences tend to be greater if you are born, speaking globally, in the West as opposed to the East, the South versus the North. They are even greater if you are born in different centuries in the same region of the planet.

The Sociology of Free Will and Consciousness

The illusion of a free-willing individual self is fueled by our self-awareness. But self-awareness is not there at birth; it develops with experience. It also has a development at the collective level as a result of cultural evolution. The opportunities for self-awareness and the experience of free will increase with an increase in the scale and complexity of social systems. Prior to the emergence of the social sciences in the nineteenth century, concepts of the self were rooted in metaphysics and theology. They supported the idea of a unique and ineffable soul at the core of each individual. The natural sciences supported a concept of the self as a collection of somative sensations. Psychology was fueled by these traditions of the person as an individual rather than a social entity. Even where psychologists recognized the significance of social factors for understanding the self, they did not attend to the relationships between self and society or view the self as a social product.

In the early decades of the twentieth century, a second generation of sociologists synthesized the writings on the social self that had emerged in the years between 1840 and 1920. They paved the way for a view of the self as a social product and social process. The concepts of self, mind, and consciousness became crystallized as the products of a social process. Self-perception and meaning are deeply rooted in the praxis of subjects engaged in social encounters. The mystery of consciousness has been fueled on the one hand by the Procrustean efforts of physical and natural scientists and philosophers and on the other by ignoring the discoveries of social scientists. The scientists and philosophers have assumed that the secrets of consciousness were hidden in the brain; social scientists finally recognized something Nietzsche and Marx among others had already noted: consciousness is a social phenomenon, a relational network, a *social* fact.

Existence in a community precedes the emergence of individual consciousness. Our participation in different social roles and positions is the foundation for our consciousness of self. This process echoes (1) the discoveries by earlier sociologists, notably Emile Durkheim (1858–1917), Ludwig Gumplowicz (1882–1909), and G.H. Mead (1863–1931) on individuation, the social process of becoming an individual, and (2) the evolutionary sociology of humans as always, already, and everywhere social and the most radically social of the social species.

Constructing Character

The self is neither the result of having a body, or a soul, or a fact of birth itself. It is rather in dynamic terms a social process. In static terms the self is a social structure, a set of social relationships. An easy way to see why this must be the case is to imagine what you must do if you want to get to know someone. You can't just look at someone and form an adequate impression of his or her character or personality. We will inevitably have "first impressions" when meeting someone for the first time. There are aspects of their bearing and dress that we can take as initial clues about their background and character. But here already we are pinning socially grounded views onto them, drawing, for example, tentative conclusions about their social class background and education from their posture and the way they dress and speak. This is easier in some cases than others. People might have trouble forming a first impression of social types they are not familiar with, types that are not part of their everyday environments. Older persons of today might have trouble forming first impressions of, for example, hipsters; people with limited experience of the world might have trouble forming first impressions of people in drag, or people from other cultures wearing clothing native to their countries of origin. This doesn't mean such people wouldn't form first impressions at all but their impressions would be fuzzier and not based on a great deal of past experience. All they can do is project the social and cultural types they are familiar with onto these strangers. They might indeed find their first impression efforts blocked by the cultural uniqueness of the person they are encountering.

If we want to go beyond first impressions, we have to ask questions and the questions we ask will elicit replies that place the person into social categories. Asking questions about a person's name, birth place, education, work, parents, siblings, likes and dislikes, and relationships will build a matrix of persons, things, and ideas that will anchor the person before you and build a character portrait. This matrix IS the self. It changes as our lives unfold and people, things, and ideas are added to, eliminated from, faded, and made more prominent. This process can be altered in a psycho-emotional direction in a culture dominated by the myth of individualism. In that case, the "getting to know you" questions will tend to be more about how you feel about things and what your emotional state is. The answers to such questions are not good guides to the social structure of the self which is more closely tied to how we will behave, think, and feel in given circumstances. Emotional profiles can of course be guides to social and cultural backgrounds, but they are more difficult to fix than structural characteristics. Experts in the sociology and anthropology of emotions will have an advantage over others in being able to place people socially by accessing their emotional profiles.

The Self-Matrix

In order to grasp the idea of the self as a social structure, imagine the self as a matrix—or more colloquially, a box—traveling through time and space containing at any given moment all of the ideas, people, and things it has encountered, arranged in a network that places them in more or less central and peripheral positions within the self's network. As the box-self moves through time and space encountering new environments and experiences, the elements in the box will shift their central and peripheral positions; some will become more central, some more peripheral, some may drop out of the box and become lost, forgotten memories. To build on this metaphor, the box can be thought of as having a screened off area behind which things, people, and ideas as memories can fall. This allows for the possibility of retrieving lost memories by way of hypnosis or psychotherapies. Changes in the brain can also be implicated in losing and gaining back memories due to aging or injury. At the same time, new elements will enter the box, altering its structure as well as particular preexisting elements. The social network of the box-self will not change much over time if the person does not move about the world much; it will experience substantial changes if the person travels widely, studies a variety of subjects, and generally adopts a more global as opposed to local lifestyle.

The Recurrence Theorem

In order to bring this idea home for yourself, in order to appreciate the idea that the self is socially situated, think about how your behavior changes when you move from a dorm to your parent's home, or from the playground to church. Are you the same person, judged by your behavior, including language and posture, when you engage your parents, your friends, your teachers, and your lovers? The continuity of your self is sustained by the fact that there are links and continuities among the different situations you move through as your life unfolds. The process of the self is the movement and change of the self-matrix (more informally, the box-self) over time. Continuities in context sustain a level of stability in your self-concept that gives you a sense of an essential core that is "you," that is your essence.

Contexts recur (shopping in one supermarket in your town is not much different from shopping in other supermarkets, even some in other countries). Structural recurrence refers to the general similarities across buildings, bridges, and other material structures; if you want to enter an office building, a school, a home, or a hospital you look for a door. You can expect to find aisles and hallways, stairs, escalators or elevators, doors, and windows. Temporal recurrence refers to the similar routines of time that apply to work, school, and home; how many hours do we work, how many hours do we spend in school, what times do we have our meals at home.

Environmental recurrence refers to the relative stability of our environments day to day, hour to hour, and minute to minute. Imagine if there was an earthquake every 15 minutes, or a tornado every hour. The stability of our selves is a function of the stability of our environments and of the recurrences that we encounter as we go about our lives.

As you travel around locally or internationally, you should take notice of the way your everyday environment is repeated. Your child's school is not so different from the school you went to. Even in most cases where your children go to better or worse schools than you did, there will be recognizable architectures. Your aunt's home is not so different from yours. You can navigate airports in all parts of the world with relative ease. It's not much different to take the subway in Paris, or London, or Shanghai, or New York even with the language barriers and technological differences that may confront us. This may vary across class, ethnicity, sex, and gender and the context from the perspective of specific individuals. But without some level of recognizable contextual, structural, temporal, and environmental recurrence, unaided mobility would be impossible. Indeed, in some cases it's easier for an American to take subways abroad than in New York City because they tend to be more traveler friendly in terms of signing and information centers. Life that isn't rooted in one home, neighborhood, or locale is made possible by spatial and temporal recurrence.

The Recurrence Theorem II: The Self from the Inside-Out

I have just outlined a view of the social self from the outside in. What do things look like from the inside-out, from the perspective and experience of the person in a modern society? The child's self-image and self-concept is dominated by the influences of the adults he or she encounters and is experienced as their voices, gestures, images, and emotional tones and colors. In addition, the child will internalize many religious and national symbols such as God and flag. This is more complicated for the modern child whose environment contains a great variety of media influences and technological gadgets and inputs. The general process of a consciousness populated by normatively valenced images is basically the same across space, time, history, and culture. These will come to mind as the child goes about "making choices" and those choices will be guided by prominent internal experiences of the voices of adults and the imagining of images. These voices and images will guide the child's choices and especially choices that have a moral dimension.

One can imagine a populated head, with puppet like images popping up at critical moments to guide the person's behavior. You might wake up even in your teens and think, "Should I go to class or skip class and go to the beach?" There will come granny's voice or your dad's telling you to go to school! The sign of a mature self-concept is the homogenization of this population so that while its influence is never eliminated its homogeneity is experienced as free will. The loss of distinctions in the populated head as a result of maturity and education is the source of the illusion of free will. The person's behavior continues to be guided by his or her cultural programming.

The process is more complicated. People in impoverished communities may not experience a sociologically sophisticated awareness of their selves but they may be more viscerally aware of the social forces impinging on their lives and consciousness. I don't mean to restrict their awareness to "gut feelings"; their experiences may train and educate them to a deep and profound political realism. They may indeed have a more realistic understanding of how the structures of power and influence in the society at large work than their more privileged peers. The developmental phase of becoming a self is in principle a life-long process. It will last longer to the extent that a person lives an active life in which he or she is continually engaging new things, people, and ideas in the world at large. One's life can in some instances become more contained and restrictive and in such cases the developmental phase will slow or for all practical purposes stop. Development necessarily involves engaging new things in the world. The continuity of the self is a function of the continuity of the person's social, cultural, and material environments. A certain degree of constancy and persistence in one's social and natural environments is a necessary condition for the continuity and stability of the self. The unity and structure of the self always reflects the unity and structure of those environments. A certain amount of disunity and chaos in those environments can be tolerated but at some point that disunity and chaos can reach a tipping point and disrupt the person's equilibrium to the point of leading to physical, emotional, and mental instabilities. No environment is perfectly stable so even in the best of circumstances the self is changing imperceptibly from moment to moment. The stability of the self depends on an environment that is stable for all practical purposes. People's sense or construction of self will be volatile and vulnerable if they grow up in volatile and vulnerable environments. If we had sensitive enough instruments, we would notice that our posture is in constant rhythmic motion in relation to its immediate environs. Everything we view impacts our posture. Our postures resonate with the objects and ideas we engage. It changes as we shift our gaze from a Picasso to a Dali, from listening to Mozart to listening to Elvis Presley. This is the "dance of the self."

Contexts, situations, structures, and symbols all have to varying degrees recurring properties. Recurrence is more noticeable if you are from a major urban area and highly mobile than from a small village in which your mobility is restricted. Recurrences make worldwide mobility possible and other things being equal we can move about without becoming complete strangers once we leave our local milieu. Our mobility is lubricated by these recurrences.

What we think and do is a consequence of our movements through social, cultural, and environmental spaces. Watch the way ideas appear to "come to you" unbidden. We don't think so much as we become aware of thoughts generated by the interactions between our apparatus of consciousness and the affordances of our environs. Consciousness emerges and is sustained at the nexus of the flow or stream of affordances and the flow or stream of consciousness. I discuss affordances in more detail in Chap. 4.

The Twentieth-Century Self

Social worlds vary across time, space, and culture and thus there are a great variety of selves that are possible in cultures and within an individual life. The large-scale changes and worldwide disruptions of the twentieth century made it clear that the unity of the self could not be taken for granted. These changes, manifested in such phenomena as alienation and dehumanization, have made it clear that the self is problematic. The digitalization and informaticization of the world have complicated this process and made the self even more problematic. We operate in a world of online and virtual selves co-existing with our everyday multiple selves. The self is constantly in motion as an agent of, in resistance to, and as a passive actor in the crucible of social and environmental changes. It is programmed by a complex set of interactions that link genes, neurons, social, and environmental interactions.

Open and Closed Systems

Programming can be more or less open or closed, thinking here in terms of open and closed systems. All systems, human, natural, and physical are more or less open; there are no perfectly closed systems in the real world. By definition, open systems exchange information and resources across their boundaries; closed systems cannot exchange information and resources across their boundaries. Open system programming gives us behaviors that are more or less unpredictable. Large, complex open systems can still yield patterns and probabilities but the analyses require greater and greater dependence on multivariate analyses and computer time. Predictions are only possible in systems that approximate closure. Open systems eliminate determinism for those who worry that I am suggesting we are or can be mechanical or robot like in our behavior. But they do not eliminate lawful behavior.

The Programming Problem

There is a large literature on the human-machine or human-computer analogy. The strong analogy is based on the argument that a finite deterministic automaton could in principle perform all normal human functions. Theologically inclined opponents would argue that humans, unlike machines and perhaps other animals, have souls. More secular opponents would ground human uniqueness in ideas about mind and free will. Those in this camp might ask what it could possibly mean to speak of a machine that is conscious, understands risk and trust, and exhibits courage, understanding, overcoming, and the capacity to worship. Programming advocates might argue that we are, in fact, machines. We are, indeed, organic networked machines, just the kind of machines that can exhibit consciousness, emotions, and the capacity to worship. This then would not preclude the possibility of developing mechanical machines capable of exhibiting their own versions of consciousness and emotions. It is important not to get trapped here by the idea of organic machines as free-standing, autonomous machines. Their—our—abilities and propensities are socially grounded. Furthermore, we don't want to get trapped by claims that mechanical machines can't think and feel the way we organic machines do. If my view is correct, mechanical machines properly conceived, manufactured, and networked will have their own versions of feelings and consciousness. If we understand culture as a speciating mechanism, we can better understand why it is often difficult to communicate with and understand members of our own biological species. That difficulty may be even more complicated when the other cultural species is mechanical.

The value of the programming analogy is that it helps us to see that the variety of human experiences can be more or less expansive, more or less narrow, more or less open, or closed. Some experiences will help to sustain a thread of continuity that accompanies changes in the self. Extreme experiences, such as war, imprisonment, and life in a concentration camp can break that thread of continuity and severely damage the self. Thanks in part to a world of never ending local wars and skirmishes, post-traumatic stress disorder is now an everyday widespread experience that brings this kind of extreme experience into the mundane prosaic world. The dependence of the self on relatively stable social relationships is dramatically exhibited in cases of human isolation.

Experiments in sensory deprivation can quickly damage the sense of self and could in principle if carried to unethical and in-human extremes lead to severe cognitive impairment and death. Such extremes are not reached in the laboratory but do so in real life "natural experiments," some as extreme as holocausts. Numerous cases of social isolation are documented in the literature and include the highly publicized cases of Kaspar Hauser, the Indian children Amala and Kamala, Victor of Aveyron, and Anna and Isabelle. Early on such cases fueled ideas about "wolf children" or "feral children." But in every case the children were victims of severe isolation and various levels of abuse by family members; they were not "raised" by wolves or other animals. We can also turn to recent cases of dramatic cultural change leading to devastating changes in the collective lives of human beings. We can see now theoretically why one of the most damaging things you can do to a prisoner is to put him or her in solitary confinement. Placing prisoners in isolation should be considered cruel and unusual punishment.

Evolution and the Self Revisited

Evolution has made us the most social of the eusocial species. We are a species that is always, already, and everywhere social. Special circumstances of history, society, culture, and biography must be present to facilitate the process of individuation that can give us a sense of an autonomous free-willing agency. But the fact of the matter is that our individuality, while real, is defined by the particular groups we have encountered in the unfolding of our lives.

There is an evolutionary context for the social self. This was the fundamental discovery of the social scientists and social philosophers of the nineteenth century who crystallized the social sciences. Their view of the social self persistently overshadowed by the myth of the individual has had to be rediscovered again and again in the wake of their initial discoveries. E.O. Wilson is the latest rediscoverer. The prominent Pulitzer Prize biologist unlike many rediscoverers is aware that he is standing on the shoulders of giants cum social networks. He doesn't directly mention the sociological roots of the idea but he is conversant with its anthropological and social psychological roots. The fact is that few if any social scientists have done as well as Wilson and for such a large audience in articulating the evolutionary origins of the social and the uniqueness of human social life. His 2012 *New York Times* best seller *The Social Conquest of Earth* presented a powerful antidote to the myth of individualism but it is not likely that it impressed readers in a way that would provoke them to abandon that myth.

Wilson is not at the end of the day able to transcend his biological disciplinary roots and use his social insights to explain human behavior. He tackles religion and falls back on biology to explain religion in what he has already identified as a radically social species. The invisibility of sociology in these instants is an interesting problem in the sociology of knowledge and not simply a matter of disciplinary blindness—I will discuss this further on in terms of my theory of social blindness.

From a biological perspective humans belong to the general category of "eusocial" species. In eusocial species, as we have seen, group members exist in a generational framework and are primed by evolution to cooperate in more or less complex divisions of labor. In this sort of discussion biologists are inclined to focus on altruism. Altruism is a psychological and individualistic concept that draws attention to the individual motivations behind our moral actions. It is more accurate to think in terms of a "cooperative imperative" for a social species. Wilson is very clear about the ways in which humans are socially different from the ways other animals are social. Humans stand apart as a social species because they are reproductively competitive (an assumption that conflates evolutionary, biological, and social perspectives and needs further consideration), and they are able to form alliances, cooperative and communication networks and mechanisms within and across groups that are the most complex among the eusocial animals by orders of magnitude.

Certain biological proclivities in our earliest ancestors provided the scaffolding for the emergence and sustained development and evolution of humans as a social species. These include large size, limited mobility, bipedalism, and among the primates specialization for life in the trees. Other key pre-adaptations include grasping behavior, the great and opposable thumbs, the development of nails from finger and toe tips, cutaneous ridges on the palms and soles, enhancement of the sense of touch, increasing dependence on vision, binocular vision, longer and straighter legs, longer feet, a reshaped pelvic region, flexibility of the forelimbs, enhancement of the ability to throw things and kill at a distance, evolution in the "complex mosaic" of local habitats that was the savanna forest, control of fire, the campsite, and larger brains. A key development in the evolution of the eusocial species was the nest. Eusocial species build nests which turn out to be a defense against intruders. The nest also sets boundaries that in a very general way define in and out groups. The campsite is the functional equivalent of the nest among humans. Wilson tells us something sociologists and anthropologists have known virtually from their beginnings, that the evolution of society has been driven by culture not by genes. There is a growing appreciation for this fact but it has not penetrated deeply into the sciences and certainly not into the popular imagination.

Just as in the case of evolution and natural selection in general, the evolution of society is the result of a series of random and chance pre-adaptations. These pre-adaptations do not follow a goal-oriented blue print. They are adaptations in their own right and operate in accordance with the general principle of natural selection. It is the failure to understand the science of pre-adaptations that leads intelligent design (ID) and creationist advocates to believe they have found a fatal flaw in evolutionary theory. Their idea—irreducible complexity—is that biological systems cannot evolve by passing through successive small modifications operated on by natural selection. The other "alternative fact" they offer to evolutionary theory is known as "specified complexity." Specified complexity is the basic sign of intelligent design. The signature event is contingent and not easily repeatable by chance; in brief such an event exhibits an "independently given pattern." ID and creationist advocates have no understanding of the process of science or the nature of theory. Without this understanding, evolutionary theory will forever elude them. One serious consequence of this is the perpetuation of ignorance within the ID and creationist communities and aligned believers. This is the sort of thinking that leads to the necessity for "Marches for Science" instead of taking a critical appreciation for and support of science for granted.

Given the evolution of eusocial species and in particular the evolution of humans we are led to conclude that it is our command of social skills not "general intelligence" that gives us our evolutionary advantages. We were never as Wilson assumes unorganized field wanderers who evolved into cohesive "campsite carnivores." This is a version of Hobbes' "war of all against all" followed by a "social contract" mythology. We begin as more or less and variously organized field wanderers. As we evolved our networks became increasingly complex, dense, and settled, a symbolic life and eventually language emerged, and all of this enhanced our species' innate musicality and rhythmicity. We became the most rhythmic of the social animals. This is significant because rhythmicity is the foundation for imitation, communication, consciousness, and compassion and empathy. We have been seriously distracted by physicists and biologists who have convinced us that we should look for the origins of consciousness in the autonomous evolution of the brain, neurons, or the quantum mechanics of the brain. The "mystery" of consciousness was already essentially solved in the early decades of the twentieth century and adumbrated in the writings of the pioneer sociologists of the nineteenth century. The mystery is solved by turning from the toolbox of the physical and natural sciences to the toolbox of the social sciences.

Wilson joins an increasing number of non-social scientists who are providing the theory and documentation for the concept of gene-culture co-evolution. This is the root idea for my model of the social brain and a new understanding of socialization. The unit of socialization is not the human being as an entity that society operates on. The unit of socialization is the system of interconnected subsystems and elements from molecules and cells to genes and neurons, neural networks, the brain and the other organs of the body, and the body coupled to society, culture, and the environment. The causal flows in this process vary in intensity and form depending on the systems and elements involved. But all are acted on simultaneously by the socialization process. I discuss the details of this model in Chap. 5.

Two factors need to be added to this introductory picture. The first is that culture is a speciating mechanism. I discuss this further on. The second is that the coupling that links biology and culture in evolution is fragmented. There are various ways in which cultural evolution outpaces biological evolution and de-couples from strict co-evolution. In strict co-evolution, we would expect a close correspondence between biology, culture, and environment. Fragmented co-evolution helps explain why humans are capable of living in ways that destroy the environment they depend on for survival. Culture is not only a speciating mechanism; it is an incredibly efficient system for destroying planetary ecologies.

The Invisible Social

As we try to understand Einstein as a social being and not an independent, isolated "I," "soul," "self," or "individual" it is important to be aware of the barriers that block this understanding. The most important barrier is what I have identified as and named "dissocism." Just as there is an autism spectrum disorder which can be described as mind blindness (Baron-Cohen 1997) there is a dissocism spectrum disorder which can be described as social blindness. Dissocism is the inability to "see" the social, that is, the inability to see the causal influences of the social life we are all imbedded in. We see people, we know they affect us in various ways but we do not see them as the basis for a collective causal impact on the very nature of who and what we are, what we do, what we feel, and what we think literally moment to moment and over the long term.

Social blindness is not due to ignorance; it is programmed and stronger the greater the emphasis in any given society on individualism. This can be intensified by religious beliefs about salvation, souls, and survival here and in the hereafter. Wilson is on the dissocism spectrum. This is a common affliction among physical and natural scientists, philosophers and psychologists. Wilson, like psychologist Michael Gazzaniga, suffers from a relatively mild case that allows him to recognize social influences but prevents him from seeing those influences at work in the case of social institutions. Both Wilson and Gazzaniga lose track of their sociological insights when it comes to explaining religion; here they see no explanatory options but biological ones.

For all of his insights into the radically social nature of human beings Wilson resorts to biologistic reasoning when he tries to unravel the origins of religion. This is exactly the same trap that Gazzaniga falls into. Wilson recognizes that there is a relationship between religion and tribalism. At this point he is on the threshold of re-discovering the classical sociological theory of the origins of religion in the rituals of our earliest ancestors. Wilson also understands that compassion which is widely and correctly associated with religion is a cultural universal. What he, along with Karen

Armstrong, the Dalai Lama, Desmond Tutu and many figures from the world of entertainment, the arts, and politics (see Armstrong's "Charter for Compassion") miss is that compassion is a centripetal force. Philosopher Peter Kivy has a much more severe case of dissocism as I show by reviewing his analysis of sociologist Tia DeNora's social constructionist view of Beethoven's "genius" in Chap. 4.

The Nature and Limits of the Golden Rule

In its most fundamental form, the Golden Rule, compassion, seems intuitively and transparently to be a universal of human life. The problem is that it works centripetally; in any given social group it's arrow of operation is inward not outward. It is therefore not conducive to cross-society, cross-cultural connections, communication, and cooperation. It can and does lubricate the levels of cooperation and communication that we are able to achieve in practice across social and cultural boundaries. But because it is a centripetal force there are significant limits to what we can achieve readily and immediately. On the hopeful side, the universality of compassion does have long term promise for achieving the goal of the Charter for Compassion, to have the peoples of the world embrace the core value of compassion across all cultural, religious, ethnic, and other social barriers.

Wilson and Gazzaniga let their biological training and education blind them to the sociological insights on society as sui generis, on God as the reification of a society's view of itself, and on religion as the institutionalization of God as a representation of the society, of the tribe itself. Wilson understands religion in terms of a biological logic. He doesn't have a well-ordered theory, only a pathway to an explanation. From his perspective the only things that can account for religion and in particular for the "phantasmagoric" nature of creation myths are physical laws, or physiology and medicine. It is no surprise that as a biologist he dismisses physical laws, totally ignores the possibility of social laws, and turns to the biologist's default explanatory framework, physiology, and medicine. The result is, as in all cases where physicists, biologists, and chemists (never mind philosophers and theologians) attempt to explain social facts a tortured and in the end unsatisfying explanation. This is the only place such explanations can end up without the use of our knowledge of symbols, language, and ritual.

Conclusion

By the time humans arrive, evolution has already made cooperative action a survival mechanism. The process begins with the increased adaptive potential of multi-cellular animals over single cell animals. This cooperative principle in evolution eventually leads to the socially dense and complex networks of human societies. I am pointing to something much less counter-intuitive than it may at first appear. Imagine a world in which individuals are the loci of material and symbolic production; this is as absurd as imagining a world in which language develops without people living and talking together. It would have been more difficult for the myth of individualism to take root if more attention had been devoted to the central part the cooperative principle plays in Darwin's theory as opposed to focusing on "survival of the fittest." That phrase does not come from Darwin; it belongs to Herbert Spencer. The idea that competition is central to evolution and progress has become part of our cultural folklore (even in science to some extent) and comes to us through the efforts of the Social Darwinists.

There is some question about the origins and emergence in the popular consciousness of the term "Social Darwinism." It appears to have been introduced by Joseph Fisher in an article on landholding in Ireland in 1877. The historian Richard Hofstadter is credited with introducing the term and seeding it into the language of scholars and the lay public by way of the 1944 publication of his book, *Social Darwinism in American Thought, 1860–1915*. It is a difficult idea to pin down in many respects but generally refers to efforts to apply ideas wrongly attributed to Darwin concerning the virtues of competition and individualism in social evolution. It has tended to support unregulated competition, imperialism, and racism. It can be thought of as the ideology of the Robber Barons, but this idea has to contend with the fact that men like Andrew Carnegie were great philanthropists. Robber Baron behavior and philanthropy are not incompatible with an economy trying to model itself on some version of capitalism. This multiplies the number of myths we have to contend with in what looks like a simple matter of brains and humans. I have dealt with the myth of capitalism elsewhere (Restivo 2018) and so bracket it here in order to stay focused on Einstein and his brain.

The literary theorist and sociologist Roland Barthes wrote that Einstein's brain had become a mythical object. It is clear that Einstein himself has become a mythical object, an idea that becomes even more powerful if we believe that Einstein was his brain. Einstein as icon is already a complicated figure. He is, Barthes writes, simultaneously a magician and machine. He is variously a comic Chaplinesque figure, a mad professor, and a humanist concerned with peace, internationalism, and co-existence. He is the sage who can be consulted on everything from world peace to the nature of God. Setting him apart as an unapproachable genius has competed in recent times with efforts to discover the human being behind the icon. We have learned that the inventor of relativity theory was a prankster, a man who had many affairs, and a man who loved to climb mountains and sail. None of these humanizing efforts have been strong enough to modulate the genius label that seems to be synonymous with "Einstein."

Bibliographic Notes for Chapter 2

The Eusocial Species

I follow the discussion of "eusocial species" in Wilson (2012); and see Gorney (1973).

The Sociology of Free Will and Consciousness

- The Social Self, Community, Cooperation, and Compassion
  Durkheim, E. (1995/1912). Gumplowicz (1980/1885); Mead (1934); Restivo (2018); on the argument that consciousness is a social phenomenon, see Whitehead (2008);
  On humans as the most radically social of the eusocial species, see Wilson (2012); Armstrong's *Charter for Compassion* (2008): https://charterforcompassion.org; on the origins of compassion and the need for belongingness in primate populations, see King (2007); on Social Darwinism, see Hofstadter (1944).
- The Cooperative Principle in Evolution
  Montagu, A. (1952); Gorney, R. (1972).
- The Illusion of Free Will
  The classical starting point for neuroscience research and theory on free will and the brain is the work of Benjamin Libet (2004). Libet's surprising finding was that unconscious neural processes precede conscious acts. The finding is rather more complicated than that. For a more broad-ranging discussion including critical commentaries, see Libet, Freeman, and Sutherland (1999); and see Caruso (2012). Wegner (2002) includes an extensive bibliography which includes

references to John Bargh's work on automaticity. Neuroscientist Hannah Critchlow (2019) sees two sides to the issue of free will and the brain: on the one hand, there is increasing evidence from the neuro- and life-sciences that our destinies are in our genes and neurons; on the other hand, we learn more everyday about the brain's plasticity, dynamism, and flexibility and environmental influences on the structure and function of the brain. D.F. Swaab (2014) takes a decidedly deterministic position on the proposition that we are our brains. The free will debate can be resolved by taking into account the distinction between open and closed systems and between determined and lawful relations: see Restivo (2018: 57–74; 22, 39); the distinction between determined and lawful relations is discussed in detail by Bohm (1971). And for a coherent anti-brainist understanding of neuroscience, see H. and S. Rose (2016).

We would have trouble denying that we are thermodynamic systems subject to the laws of thermodynamics, or that we are biological systems subject to the laws of biology. Denying these ideas leads to unnecessary mysteries and absurdities about how our bodies function. But we are not yet, most of us, ready to acknowledge that we are also social systems subject to the laws of sociology. We are as constrained by the laws of sociology as we are by the laws of thermodynamics. The fact that the laws of sociology are not yet firmly established in our culture doesn't change the reality of their impact on our lives. It's important when considering this idea to keep in mind the distinction between open and closed systems and the associated distinction between determinants and laws.

References

Baron-Cohen, S. 1997. *Mindblindness.* Cambridge, MA: MIT Press.

Bohm, D. 1971. *Causality and Chance in Modern Physics.* Philadelphia: University of Pennsylvania Press.

Caruso, G.D. 2012. *Free Will and Consciousness.* Lanham, MD: Lexington Books.

Critchlow, H. 2019. *The Science of Fate: Why Your Future Is More Predictable Than You Think.* Hachette, UK: Hodder & Stoughton.

Durkheim, E. 1995. *The Elementary Forms of Religious Life.* Trans. K.E. Fields. New York: The Free Press, orig. publ. in French, 1912.

Gorney, R. 1972. *The Human Agenda.* New York: Simon & Schuster.

Gumplowicz, L. 1980. *The Outlines of Sociology*. New Brunswick, NJ: Transaction Books, 1980; orig. publ. in German, 1885.
Hofstadter, R. 1944. *Social Darwinism in American Thought*. Boston: Beacon Press.
King, B. 2007. *Evolving God*. New York: Doubleday.
Libet, B., A. Freeman, and K. Sutherland. 1999. *The Volitional Brain: Toward a Neuroscience Free Will*. Exeter: Imprint Academic.
Libet, B. 2004. *Mind Time: The Temporal Factor in Consciousness*. Cambridge, MA: Harvard University Press.
Mead, G.H. 1934. *Mind, Self, & Society*. Vol. 1. Chicago: University of Chicago Press.
Montagu, A. 1952. *Darwin: Competition and Cooperation*. New York: Henry Schuman.
Restivo, S. 2018. *The Age of the Social: The Discovery of Society and the Ascendance of a New Episteme*. New York: Routledge.
Rose, S., and H. Rose. 2016. *Can Neuroscience Change Our Minds?* Cambridge: Polity Press.
Swaab, D.F. 2014. *We Are Our Brains: A Neurobiography of the Brain, From the Womb to Alzheimer's*. New York: Spiegel & Grau.
Wegner, D.M. 2002. *The Illusion of Conscious Will*. Cambridge, MA: Bradford Books.
Whitehead, C., ed. 2008. *The Origin of Consciousness in the Social World*. Exeter: Imprint Academic.

CHAPTER 3

Einstein's Brain: A Conspiracy of Mythologies

Abstract In this chapter, I consider the decades of research carried out on Einstein's brain and why they have proved to be, and indeed were destined to be, sterile. Intelligence, creativity, and genius are social phenomena. Einstein did stand alone and did not create ab novo. Genius is not as commonly supposed an individual attribute. It clusters, and genius clusters are associated with the rise and decline of civilizations and cultural areas. The myth of individualism supports the idea that Einstein made discoveries "merely by thinking" and that he embodied pure intellect.

The myth of Einstein as "pure intellect" is sociologically untenable and is based on the grammatical illusion of the "I" and the myth of the brain as an isolated biological organ independent of external causal factors. I introduce some of the initial elements related to the social brain paradigm such as the connectome, neuroplasticity, and regional concepts of the social brain (as opposed to wholistic concepts).

Keywords Connectome • Neuroplasticity • Einstein's brain • Studies of Einstein's brain

In this chapter, I consider the decades of research carried out on Einstein's brain and why they have proved to be, and indeed were destined to be, sterile. In Chap. 1, I wrote:

S. Restivo, *Einstein's Brain*,
https://doi.org/10.1007/978-3-030-32918-1_3

> Now we can see the mistake we—along with Einstein himself—have made in trying to understand Einstein the man. The tendency has been to view him as a unique individual and to look to genes and neurons (and more broadly biology) to explain his uniqueness. The label "genius" adds a divine factor to the explanatory narrative. I am not suggesting that we deny his uniqueness but that we revise our understanding of that uniqueness, that we contextualize it.

Intelligence, creativity, and genius are social phenomena. Einstein did not stand alone and did not stand on the shoulders of giants. He stood on the shoulders of social networks. Notice that the "giants aphorism" places the discoverer on the shoulders of a group of isolated independent giants and does not make him or her an embodiment of the network. This is true even though networks link teachers and students, collaborators, correspondents, relatives, and colleagues. Where relationships in networks are documented (as in Collins 1998), they do not sociologize the grounds for the "giants metaphor." There are networks that connect the giants but the Newton's and Darwin's are made to stand on the shoulders of individual greater and lesser giants. With the social matrix idea in our toolkit we can now see why we all stand on the shoulders of greater and lesser "giants" in social networks. In the case of Einstein, the "genius of all geniuses," we are forced first of all to focus on the very idea of genius. I do this in detail in Chap. 4.

Genius is not as commonly supposed an individual attribute. Assuming for the moment that the term "refers," genius does not appear randomly. It clusters, and genius clusters are associated with rapid changes—the rise and decline of civilizations and cultural areas. Second, Einstein's originality must be reconsidered. His originality is not as commonly supposed that of an isolated lone wolf who surprises the world with an ab novo discovery. Unexamined appearances notwithstanding, he did not make discoveries "merely by thinking"; he was not, as Golden (1999) claimed, "the embodiment of pure intellect." On the individual level, the body cannot be left behind. Socially and historically, inventions and discoveries more often than not appear as multiples. The same or similar inventions and discoveries at various stages of development appear simultaneously within the same cultural orbit.

As a member of the human species Einstein shares in that specie's unique nature as the most social of the social species. Humans do not come onto the evolutionary stage as individuals who then become social. They arrive already, always, and everywhere social. The "I" is a grammatical illusion;

the individual as an entity is not shaped from the inside-out by genes and neurons (or from a theological and philosophical perspective, by a soul and free will). This is the myth of individualism. We become individuals from the outside-in, shaped by society, culture, and our environment. Genes and neurons are not irrelevant, but they need to be fitted into a new outside-in narrative and as we will see later a more complex outside-in/inside-out dialectical process. In more contemporary terms, this process will come into view as a circulation of information driving a connectome. The connectome is the "wiring diagram" of the brain; in Chap. 5 I will expand it to connect brain, culture, and world. To anticipate, I want to expand this idea in two directions: first, I think we can imagine a micro-connectome that can be applied to subsystems of the brain and body and to micro-micro systems at the cellular and molecular levels; and second, we can imagine macro-connectomes at the levels of society, culture, and environment. The General Connectome would be the wiring diagram of the entire global network traversed by the circulation of information.

> Einstein's uniqueness is real but it is a socially grounded uniqueness.

Einstein and Einstein's brain have become mythical objects. His brain is the Holy Grail of (neuro)science. Just as the ancients transformed prominent people into gods when they died (Plato, for one prominent example), we have transformed Einstein into a genius, the genius of geniuses, a God of the gods. The term genius is charged with its original meaning of a deity within, so it's not too much of a stretch to equate "genius" and "god." Einstein and his brain have been placed in and protected by an iron cage which allows us to worship and be awestruck but not to inquire. Traditional scientists have not been able to open this iron cage. The key lies in the hands of the social scientists.

There is no Einstein per se just as there is no you or me per se. The idea that we are individuals isolated from and independent of social and cultural forces is a myth. I cannot repeat this "mantra" too many times: the human species doesn't appear on the evolutionary stage as a collection of individuals who then become social; we come on stage already, always, and everywhere social. The reason this mantra needs to be repeated again and again and its message shown to be grounded in decades of social research is dramatically illustrated by the recent special issue of *National Geographic* on genius discussed in Chap. 1.

If, as I argue, Einstein had more than the force of his "own" thoughts to draw on for his theories, what is the rationale for assuming a post-mortem analysis of his brain can identify innate genetic and neuronal factors peculiar to Einstein? If Einstein is a grammatical illusion, how can we hope to find "his own thoughts" in the architecture of his brain? How can we hope to find the thoughts of his social self in that architecture? The "always, already, and everywhere social" thesis is the basis for an alternative narrative. I assume that the differences that have indeed been found in some of the comparative research on Einstein's brain using small samples of "normal" brains reflect his life experiences. The rationale for this alternative narrative goes beyond a simple social theory. It is grounded further by the research on brain plasticity, epigenetics, and the known ways in which environmental conditions influence brain structures and processes. My objective in this book is not to deny Einstein's uniqueness but to demonstrate that it is a socially grounded uniqueness, a function of the unfolding of his life as a process of moving through a unique series of social influences. Where in this idea of Einstein's "own" thoughts is there room for his intimate associations in the special relativity years with his wife Mileva (independently of whether or not her contributions have been exaggerated) and scientists he knew personally or through their writings; and later with Besso and especially Grossman who helped Einstein with the non-Euclidean ideas and the concept of tensors he needed to formulate the general theory?

The findings on brain plasticity have helped to undermine the idea of a completely biological brain. Brain structures and processes are influenced by social, cultural, and environmental forces (see the section on "Environmental Enrichment/Enhancement" in the notes). We have learned much within the last 70 years or so about brain plasticity and environmental influences on brain structures and processes. More recently it has become possible to sort out social and cultural influences from general environmental ones and this has led to the idea that brains are constitutively social and cultural. This can be true without the need to radically eliminate genetic and neuronal influences. Beginning as early as the 1950s, the concept of a social brain arose, evolved, and crystallized when the psychiatrist and neuroscientist Leslie Brothers introduced the concept of a social brain into the core neuroscience literature in 1990. She proposed a regional theory; that is, she identified three specific regions of the brain that constituted the social brain. Since then a more wholistic view has emerged (see Chap. 5).

In the social brain paradigm, the whole brain is viewed as a social thing. In a more sophisticated vein, I argue that the brain should be viewed as (1) a social thing all aspects and parts of which are socially and culturally constituted; this should not be construed to mean that the brain is social per se; genes and neurons play a role but that role is complexly interconnected with social, cultural, and environmental influences; (2) the material site on which neuroscience, life science, and social science forces converge, intersect, and interact; and (3) at the most general level, a level that erases the disciplinary disciplining of the brain, a node in an information network of entities from cells to genes and neurons, from neural nets and organs to the central nervous system, and from the body to society and environment that constitute the thinking, acting, conscious human being.

My interest in the brain began about the time that President George Bush proclaimed the 1990s "The Decade of the Brain." In December of 2010, I attended a conference at Oxford University on the rise of the brain as a focus of scientific and public interest and the emergence of a brain industry. These two events anchor a period during which the attention space devoted to brain research reached a critical level of public, media, and scientific awareness. My work contributes to documenting this development and tells the story in a way that draws sociology into the center of the conversation. There are no isolated minds or brains in vats. The brain is an active organ and whatever we mean by mind arises in social networks. A recent article on the mind mystery (Bertolero and Bassett 2019: 26, 33) focuses on the "carefully orchestrated interaction among different brain areas." In their concluding paragraphs, the authors note that "environment" might have something to do with conditioning the structure of the brain and adding functional capacities. We see that in this very moment, the brain-centrists still have no pathway from such minor insights to the nearly 200 years of speculation, theory, and research from Durkheim, Marx, and Nietzsche to James, Dewey, and Mead and from Mead to C. Wright Mills, Randall Collins, and me that offers a sociological answer to the "mystery" of mind.

Rationale

I have been motivated in undertaking the sociological study of mind, brain, and genius by "invitations" from leading figures in the philosophy of mind, neuroscience, and cognitive science. Philosophers have come to recognize the special role social life plays in structuring consciousness without understanding that role (Searle 1992); neuroscientists increasingly argue that we must take social and cultural content into account when studying the brain but they don't know how to do this (Damasio 1994); and cognitivists have been forced to admit that cultural factors cannot be ignored in mind and brain studies, but they have nothing substantive to say about this (Franklin 1995). We are seeing the beginnings of a readiness to admit sociology and anthropology into the matrix of disciplines studying brain and mind. And even artificial intelligence researchers and robotics engineers, frustrated by decades of unsuccessfully trying to brute force program consciousness and thinking into machines, have been grudgingly coming around to the idea that they might have to take social science and the embodiment movement into account (Restivo 2005/2019). Already in the first edition of his Sociological Insight (Collins 1982; see second edition, 1992: 155–184), Randall Collins had explained why sociologists would have to play a major role in developing any "real" artificial intelligence. The problem is that this increasing awareness isn't associated with appropriate disciplinary tools and theories. The social project therefore tends to seem too complex and daunting to incorporate. Nonetheless, some scientists and philosophers driven by ignorance and arrogance believe they should be able to master "the social" without any special training or education because the social is wrongly assumed to be transparent and intuitively accessible. The reasons for the extraordinary and wrong-headed attention to Einstein's brain is grounded in the prevailing scientific and popular view voiced in the first paragraph of George Bush's Presidential Proclamation 6158 of July 17, 1990, proclaiming the Decade of the Brain. The reason for the argument that more resources should be devoted to neuroscience research was based on the view that the brain is the "seat of human intelligence, interpreter of senses, and controller of movement."

The current landscape is dominated by two basic paradigms: (1) you are your brain, represented by D.F. Swaab's (2014) neurobiography of the brain from womb to Alzheimer's; and (2) you are not your brain, represented most prominently in the works of philosophers Alva Noë (2010) and Andy Clark (1998). The conflict is over whether we are brains in a vat or whether we are at the center of a network linking brain, body, and world. These positions, however, are two sides of the same coin; biology and environmental science dominate in both spheres and social science remains invisible.

That the "brain-in-a-vat" perspective still prevails is illustrated by the follow-up to Bush's Decade of the Brain, Obama's 2013 BRAIN initiative (Brain Research through Advancing Innovative Neurotechnology). The Obama proclamation, like the Bush proclamation, identifies the brain as the font of our thoughts and behaviors.

Classically, the brain-in-a-vat thought experiment is most commonly used to illustrate global or Cartesian skepticism. You are told to imagine the possibility that at this very moment you are actually a brain hooked up to a sophisticated computer program that can perfectly simulate experiences of the outside world. The skeptical argument, then, is that if you cannot now be sure that you are not a brain in a vat, then you cannot rule out the possibility that all of your beliefs about the external world are false. Or, to put it in terms of knowledge claims, we can construct the following skeptical argument. Let "P" stand for any belief or claim about the external world, say, that snow is white.

1. If I know that P, then I know that I am not a brain in a vat
2. I do not know that I am not a brain in a vat
3. Thus, I do not know that P.

The brain-in-a-vat argument is usually taken to be a modern version of René Descartes' argument (in the *Meditations on First Philosophy*) that centers on the possibility of an evil demon who systematically deceives us. The hypothesis is the premise behind the movie *The Matrix*, in which the entire human race has been placed into giant vats and fed a virtual reality at the hands of a malignant artificial intelligence (our own creation, of course). The "brains in a vat" idea is sometimes used to explore the idea that we could be a collection of cells in a petri dish hooked up to certain stimuli that evoke consciousness and a thoughtful awareness of a world. Hilary Putnam (1981: 1–21) and Thompson and Cosmelli (2011) have essentially erased the plausibility of the brains in a vat thought experiment. Nonetheless there is still a widespread assumption across the sciences, the neurosciences in particular, and the lay public that the brain is the essential actor in the theater of our behaviors and thoughts. The brain in this theater is a one "man" show.

The idea that we are our brains is the basis for decades of research on Einstein's brain searching for the genetic and neuronal causes of his genius. What have we learned? We haven't in my view and in the view of at least some of the students of Einstein's brain learned anything that would allow us to correlate or causally link brain architecture to genius or creativity. There do, however, appear to be differences between Einstein's brain and small samples of "normal" brains. Given the small samples available for comparison with Einstein's brain, the very idea that we have sufficient evidence to identify "the normal brain" is problematic in the extreme. With that caveat in mind, the findings on Einstein's brain have been used to support the idea that neuroanatomical features are correlated with signs and levels of intelligence.

In general, the studies of Einstein's brain show smaller regions than expected in the speech and language areas, larger regions in the case of numerical and spatial processing. Some reports show higher than normal numbers of glial cells in Einstein's brain. The research landscape on Einstein brain includes studies, contradictory findings, and criticisms. In general the criticisms are constructed within the boundaries of the brains in a vat paradigm. Harvard's embodiment of the theory of everything, Steven Pinker, commenting on one of the classic articles on Einstein's brain (Witelson et al. 1999) demonstrates the depth of this kind of thinking revealed at the policy level in the proclamations of Bush and Obama. The article, according to Pinker, demonstrated that all of our thoughts and emotions are based in the brain.

Einstein's brain was not the first subject of dead brain research. The most detailed post-mortem study of a brain was carried out by O.C. Vogt on Lenin's brain at the request of the Soviet government beginning in 1925. He was charged with carrying out a cytoarchitectonic study of the material foundations of Lenin's political genius. Vogt's major finding in the case of this "brain athlete and association giant" was extraordinarily large pyramidal neurons in the third cortical layers of several areas of the brain. But this could have been caused by neuronal swelling due to a delay in tissue fixation.

Harvey was apparently inspired by Vogt's study when he decided it was worthwhile to save Einstein's brain. He hoped that living thoughts could be found in dead tissue. If this was at all possible it was going to involve studies of the association cortices in the frontal, temporal, and parietal lobes.

Vesalius already in 1543 had initiated studies of normative neuroanatomy. The history of brain studies has been a history of studying diseased brains; by contrast, very few brains of geniuses are available. There is evidence that the Egyptians had studied the results of brain trauma as early as 3000 BCE (v. the *Edwin Smith Surgical Papyrus*). During this era, the brain was not considered an important organ of consciousness and thought; here the heart was primary. It took about 1300 years before Hippocrates and others finally identified the brain as the seat of cognition, thought, and consciousness and not the heart.

Already by the early twentieth century it was clear that the evolving science of the brain was going to be pluralistic. Vernon Mountcastle (1918–2015), the father of neuroscience, identified 19 neuroscience subdisciplines. He presided over the first national neuroscience conference in 1971; he was the first president of the newly founded Society for Neuroscience. A pluralistic neuroscience was not a promising foundation for a unitary theory of the brain.

The study of genius and the brain arguably begins with the study of the mathematician C.F. Gauss' brain (1777–1855), even though the notably "small" brain of the scientist/mathematician Pierre-Simon Laplace (1749–1825) had been studied about 30 years earlier. In the early twentieth century, E.A. Spitzka (1876–1922) studied the brains of six eminent members of the American Anthropometric Society. He is famous as an editor of early-twentieth-century editions of Gray's Anatomy (originally published in 1858) and more famous as the anatomist who in 1901 performed the autopsy on the brain of the man who assassinated President McKinley. In addition he reviewed studies of the brains of 130 "notable" men and women, among them Turgenev and Ludwig II of Bavaria. One of the brains in this group belonged to Walt Whitman but it was ruined for research when it was dropped on the laboratory floor by a research assistant. Spitzka was interested in the possible relationship between brain size and intelligence. He could find no evidence of such a relationship but nonetheless speculated that the ratio of the mid-frontal lobe to the cuneus/precuneus explained the superior conceptual powers of anatomist Joseph Leidy and the anatomist Edward Cope's abilities in abstraction. Wrong-headed as this approach was, Spitzka's focus on the importance of myelin-development and the multiple connections of nerve cells by multiple association fibers anticipated the current focus on the connectome.

Rudolf Wagner studied Gauss' brain and reported that it weighed 1492 grams and had a cerebral area of approximately 220,000 square millimeters. Highly developed convolutions in Gauss' brain were thought to be an anatomical expression of his genius. I've already referred to Vogt's study of Lenin's brain and the study of Laplace's brain; the brain of mathematician Sofia Kovalevskaya (1850–1891) was also studied. The brain of the doctor, lawyer, schoolmaster, photographer, inventor, carpet designer, phrenologist, and philologist E.H. Rulloff (1819–1871) is one of the largest ever recorded at 1673 cm^3. Rulloff was executed as a serial killer. The brain of the Yahi Native American Ishi was removed but never studied in any detail. The taking of Ishi's brain, however, is a study in the linked histories of eugenics, Social Darwinism, misguided exercises in brain science, racism and ethics in early American anthropology, and the mistreatment of native Americans and the demeaning of their culture (Starns 1994: 174–186).

Connections

The connectome can be loosely considered the wiring diagram of the brain. It is in a narrow sense a map of the brain's neural connections. More broadly, it maps all the neural connections of the nervous system. In light of the increasing awareness of the significance of the connectome it is unfortunate from the neuroist perspective that no studies have been done of the length, density, and morphology of Einstein's axons. Neuroists are scientistic neuroscientists. The term "neuroism" was introduced by Leslie Brothers (2001) to refer to the idea that the mind can be explained by the brain. Brothers' objectives were to demonstrate the logical flaws of neuroism and uncover the ways in which it naturalizes culturally derived ideas about who and what we are. In my model of the social brain, which I will introduce in Chap. 5, I will extend the connectome to encompass the information system connecting brain, body, and the social and physical environments.

At this point the connectome idea is applied within the brain. Neuroscientists have come to appreciate the fact that connectivity is the defining feature of brain architecture. This buries strict localization in the graveyard of mistaken ideas about the brain along with strict modular theories, and the left brain-right brain mythology. Cortical modules are not autonomous islands in a neurological archipelago. There is a path to the brain's coupling with society and environment in these developments; minds differ because connectomes differ. But neuroism prevails and the connectome is interpreted as more evidence that we are our brains (e.g., Seung 2012).

Harvey set out to ascertain the difference between "your" brain and the brain of a genius. He assumed that the key to Einstein's genius could be discovered using a neurohistological approach to reveal the brain's microscopic structure. The gross anatomy of the brain was not at the end of the day the royal road to genius. The last study of Einstein's neurohistology was in 2006. Since then there have been important developments in the study of astrocytes. Lepore (2018) is optimistic about the possibility that microscopists can tell us more than we now know about genius. The reason for his optimism is that as central nervous systems become more complex ("smarter") astrocytes get bigger, more complex and variable architecturally, and signal more rapidly.

Given the myth of individualism, it should not be surprising that some parties sought to obtain a little piece of Einstein's brain for DNA sequencing. Lepore describes these efforts as "mildly crackpot" requests aligned to the idea that another Einstein could be cloned. Aside from the sociologically absurd notion that Einstein or anyone else could be cloned (not simply biologically but literally reproduced as a personality) using a biotechnological "fix," Einstein's DNA sequence had been irrevocably altered by the room temperature preservation techniques used in the 1950s.

When we now refer to Einstein's brain, we are talking about a brain that was originally, following its removal, recognizable as two intact hemispheres. These were transformed by Harvey into 240 gauze wrapped pieces of cerebral tissue placed in 2 glass jars of formalin.

In 2009, Falk and Lepore re-analyzed the five grainy photos previously studied by Witelson and her colleagues. The photos showed an intact, partially dissected brain that allowed them to analyze the ridges and crevices of the cortex anatomy. Their approach was different from LaBerge's study of the microscopic connections between Einstein's neurons.

In the wake of Harvey's haphazard distribution of the Einstein brain materials, Falk and Lepore were able to access resources which by 2011 had found their way to the National Museum of Health and Medicine (NMHM). It is interesting to note that the NMHM is administered by the Department of Defense as part of the Fort Detrick complex. Falk and Lepore's analysis of the notes on and digital photographs of Einstein's brain led to their discovery of Einstein's "astonishing brain anatomy." Witelson and her colleagues had access to McMaster University's brain bank of 35 male and 56 female brains; the brains belonged to people described as having "normal" cognitive abilities. Harvey's original claim that Einstein had no parietal operculum in either hemisphere was confirmed by Witelson and her colleagues but disputed by Falk and her colleagues.

As we review these findings we notice that they depend in large part on a controversial classical localization theory that correlates neuroanatomical regions with modes of thought and cognition. For example, Einstein's lower parietal lobe was found to be 15% larger than average. This region is conjectured to be associated with mathematical thinking, visual and spatial cognition, and imagining movement.

Witelson and her colleagues reported that photographs of Einstein's brain showed an enlarged but truncated Sylvian fissure. They also reported that the parietal operculum region in the inferior frontal gyrus of the frontal lobe was vacant. The researchers speculated that the vacant areas may have improved neuronal communications in this region of the brain. The unusual fact of the missing part of the Sylvian fissure could, they speculated, explain Einstein's patterns of thinking which he himself claimed to be visual rather than verbal. Cambridge University's Laurie Hall said there was no way at the moment to demonstrate a link between the anatomy of the Sylvian fissure and particular thought processes. Such claims might be demonstrated robustly, she suggested, using neurotechnologies such as fMRIs.

Another anomaly in the anatomy of Einstein's brain was that his parietal lobe was notably larger than normal. His cortical anatomy was disorganized. Falk, Lepore, and Noe (2013) reported an unusual right frontal lobe (4 gyri instead of the "normal" 3 gyri). The extra gyrus could have reflected the expanded prefrontal association cortices according to these researchers. In fact, every lobe of the external cortex was unusual. The power of neuroistic thinking is demonstrated by the fact that in spite of any evidence for a localization theory of imagination and a global neuronal workspace, Lepore still thinks we will find a "neural network address" that localizes Einstein's creativity.

In her examination of the photos of Einstein's brain, Dean Falk saw a knob shaped fold of precentral gyrus on the surface of Einstein's right frontal lobe. This fold was not mentioned in the paper by Witelson and her colleagues. The so-called omega sign is an anatomical landmark for musicianship. This once again raised the nature-nurture question: was Einstein born with the omega sign or was it environmentally induced through Einstein's violin practice? Lepore can't resolve this problem (which of course leaves an opening for the more inviting, in my view, socio-environmental hypothesis). But here we had another example of the unusual nature of Einstein's brain, the unusual gross anatomy in and around the primary somatosensory and motor cortices. The possibility of environmental inducement is reinforced by Lepore's caution that these unusual features of Einstein's brain should be understood in terms of structural biology and not as the grounds for Einstein's genius. Localization is useful in diagnosing and treating brain diseases but not in dealing with the frontal lobe and other complex parts of the brain. Even given the "omega sign," we cannot localize where "music" lives.

Einstein's brain had a structural feature that effectively increased the amount of motor cortex devoted to the right face and tongue (see the bibliographical notes for references to Penfield's homunculus model). This expanded region bordered the diagonal sulcus in both hemispheres; this is normally a unilateral feature. The left pars triangularis was unusually convoluted compared to that on the right. This gave greater anatomical complexity to Broca's area which suggests greater language facility. This contradicts reports of Einstein's late development in language. But let me note that language development in the neurological sense is not necessarily correlated with the ability or willingness to speak. Given all of that it turns out that identifying the boundaries of the parietal lobe is more conjectural than topologically scientific.

Given all the caveats about localization and the uncertainties of topological anatomy, much was made of the anomalies of Einstein's parietal lobe revealed by the analysis of the "lost" photographs, especially the absence of the parietal opercula. Falk, Lepore, and Noe explained that the parietal opercula did in fact exist but a structural anomaly made it appear to be missing. They also discovered that Einstein's brain was not spherical. It had a frontal occipital bulge. Every lobe had unusual sulci and gyri. It is worth considering that Connolly (1950) argued that the size and gross anatomy of the brain cannot be assumed to reflect intellectual abilities. He cautioned against putting too much stock in the descriptive reports on the brains of deceased scholars.

One of the most widely reported studies of Einstein's brain was carried out in the 1980s by Marian Diamond, a professor at Berkeley. She analyzed four sections of the cortical association regions of the superior prefrontal and inferior parietal lobes in the right and left hemispheres of Einstein's brain. Harvey gave her the brain sections. She and her associates published the first study of Einstein's brain in 1985. She compared the ratio of glial cells in Einstein's brain with that of the preserved brains of 11 other males. Dr. Diamond's laboratory made thin sections of Einstein's brain, each 6 micrometers thick. She then counted cells using a microscope. Einstein's brain had more glial cells relative to neurons in all the regions she examined. This finding was, however, statistically significant only in the case of the left inferior parietal area. This area, part of the association cortex, incorporates and synthesizes information from various brain regions. We are confronted once again with a nature-nurture dilemma. Some research has suggested that a stimulating environment can increase the proportion of glial cells. Therefore the high ratio Diamond found could have been caused by Einstein's scientific lifestyle. Diamond's study was limited by the fact that she had only 1 "genius" brain to compare with the 11 normal brains she had access to.

Diamond had her critics (e.g., S.S. Kantha 1992 and Terence Hines 2014). Hines pointed out problems with observer bias. For example, the microscopists comparing Einstein's brain with controls knew which slides came from Einstein and which came from the controls. Of the 28 statistical comparisons in Diamond's study only one, the neuron-glia ratio of the left inferior parietal cortex, was significant at the p=0.05 level. We now have evidence that glial cells continually divide as we age. Einstein was 76 when he died. The 11 brains Diamond used for comparative purposes were from males 47–80 years old. All of them had died from non-neurological diseases. Once again the specter of nature versus nurture enters the picture; we have no way to control the environmental conditions brains come from. Keep in mind that chronological age cannot be trusted to be a good indicator when working with biological systems. Environments are a strong influence on organisms; human subjects are difficult to study because their environments vary and can't be controlled for.

Additionally, there is little information regarding the samples of brains that Einstein's brain was compared with. Relevant factors such as IQ scores are not known. Diamond also admitted that research disproving the findings was omitted. His brain is now at the Mütter Museum in Philadelphia and 2 of the 140 sections are on loan at the British Museum.

Diamond admitted only that it would have been better to have had eleven "Einstein" brains to study but that nonetheless she'd taken an important first step. Diamond's hypothesis was that Einstein's genius required a higher neuronal metabolism than normal. This was based on the assumption of a one to one ratio of glial cells to neurons; the control brains' ratio was one to two. The norm may be closer to 10–50 glial cells per neuron (Kandel et al. 2013) which would basically invalidate Diamond's hypothesis.

Like many of the ideas, concepts, and hypotheses surrounding Einstein's brain research, controversy has found its way into the problem of the glial/neuron ratio. Following the research that led to the standard textbook claim that there are 10–50 times more glial cells than neurons in the central nervous systems of vertebrates and Diamond's finding of a glial to neuron ratio of 0.5, Herculano-Houzel (2014) found the ratio in the whole brain to be 0.99. Her work showed that the ratio varies uniformly across brain structures and across species.

It took a little over a decade before more Einstein brain studies appeared consisting of four papers and abstracts on Einstein's microscopic neuroanatomy. It wasn't until 1999 that Einstein's gross brain anatomy came under investigation. This was in the important Lancet article by Witelson, Kigar, and Harvey (1999), sometimes referred to as Witelson (1999). Dahlia Zaidel (2001) studied Einstein's hippocampus known for its role in learning and memory. Normal brains are relatively symmetrical in terms of the neurons on the left and right sides of the hippocampus. In the case of Einstein's brain, the neurons on the left side of the hippocampus were found to be significantly larger than those on the right. The implication, according to Zaidel, was that Einstein's left brain had stronger neuronal connections between the hippocampus and the neocortex than his right brain. The neocortex is associated with logic, analysis, and innovation.

Lepore points out that Einstein had three brains (in the post-mortem context): a physiological brain, a psychological/paleoanthropological brain, and a physical brain. The nexus of these perspectives has not produced a convincing case for the neuronal basis of Einstein's achievements and indeed has opened many pathways to an alternative socio-environmental narrative of genius.

The Critics

The critics of Einstein brain studies have focused on internal issues; that is, they have criticized the studies from inside the framework of the Einstein brain studies. Selection bias has been identified as one of the factors that must be taken into account in evaluating the reliability and validity of this research. For all of the research that shows significant differences between Einstein's brain and various sets of comparison brains, there are studies that demonstrate no differences. The "difference" studies are more likely to be published due to the inertia of unexamined cultural mythologies about individuals (the myth of individualism) and brains (the brains in a vat and neuroism myths).

Some critics assume that brains are governed by a law of uniqueness: all brains are different from all other brains. This ignores patterned cultural differences as against individual differences. The claim is that given individual differences, there is no evidentiary basis for concluding that the unique characteristics of Einstein's brain explain his genius. The prospect of matching brain features to ideas, even if possible in principle, is severely hampered by the paucity of brain samples that could be used in comparative studies.

If the only tool in your tool kit is a hammer, everything looks like a nail. Sir Roger Penrose, a mathematical physicist who has walked in the shadow of the genius of Stephen Hawking as a collaborator, has one tool in his toolkit which he has decided to use to decode consciousness. That tool is quantum theory and for some reason Penrose looked at consciousness and saw a nail for his hammer. He correctly dismisses the idea that the brain is a complicated computer. The specific hammer Penrose has for unlocking the key to consciousness is quantum gravity. He believes that quantum gravity effects in the microtubules of the neuronal cytoskeleton give rise to consciousness. His opposition to the computer analogy is clearly based on his introspective self-understanding of his own complexity. He is a victim of two errors: the first is his assumption that he has direct access to the workings of his own brain or mind (thus committing the fallacy of introspective transparency); and the second is that he suffers from dissocism, leading him to view the problem of consciousness as one that he can solve with his hammer, quantum theory. Penrose embodies all of the fallacies grounded in dissocism, the myth of individualism, and neuroism.

Three of the most basic criticisms of Einstein brain studies are (1) that the brain is not the place to look for genius (dualism is wrong); (2) that if the brain was the place to look we're not smart enough to identify the signatures of genius (mysterianism); and finally (3) that the very activity of studying the brain in the search for the origins of genius is unethical (or more narrowly that the taking and study of Einstein's brain was unethical). Lepore, viewing things from the perspective of a neurologist, is skeptical about whether we will ever find a neurological basis for creativity and genius (even if one exists). On the other hand he complains about the critics of his "peer-reviewed" paper (with Falk and Noe) which presented some "fascinating and verifiable" results (the fallacy of interpretive inertia).

One of the easiest things to do when studying brains and genius is to try to determine whether genius is related to brain size. If this is done it becomes quickly apparent that the answer is no. Elephants and some cetaceans have larger brains than humans, and Einstein's brain was below average in size. Neanderthals had bigger brains than Homo sapiens. And the evolutionary record shows no correlation between increased brain size and the sophisticated use of tools. A better quantitative approach is to consider the ratio of brain size to body mass. But this fails in the end; the human brain is about 2% of body mass; it is 10% in shrews. Finally, we can consider the encephalization quotient (EQ). This measures the departure of brain size in a species from a "standard" mammal, the cat. Humans have the highest EQ at about seven and a half. The results when studying EQ among capuchin monkeys, gorillas, and chimpanzees recommend caution when using EQ as a predictor of intelligence. The EQ of gorillas, for example, is less than half that of the brown capuchin monkey. There is one more option for those looking for a neuroanatomical basis for creativity and genius; that is to count neurons and measure neuronal density. There is some evidence that suggests that humans have the highest density of neurons of all the animals.

It is no surprise that neuroist neuroscientists attribute humanity's "Great Leap Forward" 40,000 years ago to neuronal engineering. I say more about this further on. At the end of the day, the neuroist is forced to postulate as a first principle that Einstein's genius was sufficiently different from the "norm" (remember that we really don't have a good basis for establishing such a norm) that it must be reflected in the physical architecture of his brain.

Witelson hypothesized that the lack of a Sylvian fissure may have allowed the brain cells to crowd in closer to one another, which in turn enabled them to communicate much faster than normal. This brain structure may also have had something to do with Einstein's delayed speech development, which raises questions about whether it's helpful to know this sort of information about yourself if it was a possibility. If Einstein had known that his brain was different, maybe even flawed, would he have pursued academics? Galaburda (1999) argued there was no certainty that the photos used by Witelson were of Einstein's brain; and indeed there was no certainty. Galaburda and Witelson also disagreed about whether the photos showed that there were no parietal opercula. The resolution of this controversy was based on the recognition that the parietal lobe is not an autonomous anatomical region; this means furthermore that the parietal lobe cannot be identified with a specific physiological function. The parietal lobe was a topographical convenience and thus the controversy about Einstein's parietal lobe, the existence or non-existence of a parietal opercula, and whether or not Einstein was as Witelson suggested a "parietal genius" became moot.

Another controversy turns on the significance of Einstein's larger than normal corpus callosum. This structural feature was discovered by Weiwei Men (2013) using a control group of 67 younger and age matched men from the Harvey Collection at the National Museum of Health and Medicine. As a positive contributor to Einstein's genius, a larger corpus callosum signifies a greater number of axons crossing between the two hemispheres. Alternatively, there would be fewer axons if they had hypertrophied. In general, one of the overall findings of studies of Einstein's brain is that it was characterized by greater neural interconnectivity than normal; in other words his "internal wiring" was more complex and more dense (see, e.g., Lepore 2018: 1975ff.). Einstein's microscopic connectome has not yet been mapped.

At this point, scientists don't know enough about how the brain works to know if Witelson's work is accurate, though it's the going theory at the moment. For all visible purposes, Einstein's brain seems perfectly normal, if not a little damaged, with nothing that would immediately indicate any great genius. The number and density of neurons in Einstein's cortex was normal according to Witelson. What stood out for her was that the ratio of glial cells to neurons was about 1:1. Diamond's study gave the same ratio. The compactness of Einstein's supra-marginal gyrus within the inferior parietal lobule meant an unusually large area of highly integrated cortex within a functional network. This might, according to Witelson, account for Einstein's exceptional abilities in visuospatial cognition, mathematics, and imagining motions or movements.

Dahlia Zaidal of UCLA was invited by Witelson to visit her lab in 2001. Zaidal looked at the digitized slides of Einstein's hippocampus and found that cell bodies in his left hippocampus tended to be larger than the cells in his right hippocampus. She didn't think this was in any way connected to his genius. Rather she thought they reflected the normal atrophy associated with aging and disease.

Colombo's (2006) work on Einstein's brain showed nothing distinctive about the four blocks he'd studied. He saw a diseased brain rather than the brain of a genius. He criticized the earlier studies of Einstein's brain for their inconsistencies and doubted whether microscopic neuroanatomy studies could ever prove useful. One of the open questions about Einstein's brain is whether or not his brain showed signs of normal aging. The earlier findings by Diamond, Witelson, Zaidel, and others were essentially overshadowed by Colombo's studies and remarks and no studies were forthcoming for ten years in the wake of his publication. One of the latest developments fueled by recent research is an interest in Einstein's white and gray matter. Fortunately, Harvey stained these areas in 1955. While these may lead to new areas of difference between Einstein's brain and normal brains, it will not give us any new insights into the neurological grounds for his genius. It is this flawed neuroism that fuels the hope that the neuroscience of equivalent genius brains may yet yield the secrets of Einstein's brain. More generally, the neuroists expect to find that thought is produced at a certain level of cellular activity; the question is: Is it 1000 cells, 10 million cells, 100 million cells?

Harvey never gave up on his belief that Einstein's brain would reveal something special. Near the end of his life, after carting the brain around the country, he returned to the place from which he had taken it: Princeton Hospital. He gave the brain to the man who had his old pathology job. Writer Michael Paterniti, who accompanied Harvey on a cross-country trip with the brain, hypothesized in the book *Driving Mr. Albert* (2001) that Harvey picked someone who represented a sort of reincarnation of Harvey himself, something that the pathologist in question also acknowledges. "Well then, he's free now," the man told Paterniti of Harvey's choice, "and I'm shackled." If Einstein's brain ever truly reveals any secrets, Harvey won't be here to see it; he died in 2007 at the age of 94. Einstein and the mystery of his brain, however, live on.

Intimations of the Social

Lepore's *Finding Einstein's Brain* (2018) is the most recent and the most comprehensive guide to the neuroscience of Einstein's brain. It is a scientific complement to Paterniti's (2001) *Driving Mr. Albert: A Trip Across America with Einstein's Brain*, a journalistic journey. Lepore is a clinical neuro-ophthalmologist, and his approach to Einstein's brain is strictly neurological. Neither sociology nor anthropology is indexed. One of Lepore's collaborators, Dean Falk, is an anthropologist but specializes in paleoanthropology, not cultural anthropology. He does however want his readers to know that he recognizes the "profound role" culture and social interaction play in the working brain. He is certain that the brains of people during the Renaissance and the Enlightenment were anatomically the same as the brains of people in earlier generations. He cannot however see these factors as playing a role in "the Great Leap Forward" of the Upper Paleolithic age. The emergence of stone tools, body art, burial rituals, and language can only be explained by a neurologic reorganization mostly inaccessible to modern investigators. Compare the anthropologist Clifford Geertz's (1973: 76) perspective on brain and culture; the brain is "thoroughly dependent upon cultural resources for its very operation; and those resources are, consequently, not adjuncts to but constituents of, mental activity." Indeed, DeVore (Geertz 1973: 68) argued that primates literally have "social brains." The evidence for this conjecture in humans has been accumulating in recent years along with a breakdown of the brain/mind/body divisions. More specifically, Geertz argued for the synchronic emergence of an expanded forebrain among the primates, complex social organization, and at least among the post-Australopithecines tool savvy humans, institutional cultural patterns.

Increasingly we find intimations of the social in the neuroscience literature without a full appreciation of or acknowledgment of the actual and potential contributions of the social sciences. Lepore does note David Lewis Williams (2002: 96–100) cautionary note about assuming a neuronal event was behind the West's "Creative Explosion" (i.e., Great Leap Forward) 40,000 years ago. Lewis looks to "social circumstances." But Lepore is not prepared to consider this alternative non-neuronal hypothesis. Another intimation of the social occurs when Lepore points to the "fortuitous" circumstance of Einstein being raised in a "technologically adept" family. The enriched environment hypothesis is alluded to when Lepore retells the famous incident of Einstein receiving a compass; and he adds further that Einstein played with Anchor Stone Blocks with which he built complex structures. Lepore wonders if perhaps these solid geometrical toys shaped his imagination. Besides these toys and the compass, Einstein's encounter with Euclid's plane geometry had a profound effect on him. He slowly came to conclude that certain truths could be discovered by reason alone. At the same time that we are told that relativity sprang from the mind of Einstein like Athena from the mind of Zeus, we are reminded that Einstein was thinking in the context of Maxwell's equations. It was the field Zeitgeist of these equations that led Einstein to the insight that gravitation is essentially geometrical. Indeed, geometrical thinking was a feature of the cultural renaissance which gave birth to Einstein, Picasso, and the wider "genius" network of the period 1840–1920 (see Chap. 1).

Another intimation of the social shows up in Lepore's remark that given neuroplasticity we can assume that Einstein's 1905 brain was not the same as his 1955 brain. We should ask what are the fuel and motor of neuroplasticity. And indeed, Lepore points out that the dynamic wiring of the brain's neural networks is a function of nurturing influences. It is not possible to construct a mental map, whether a general map of intelligence such as Spearman's "g," Gardner's "multiple intelligences," or Freud's id/ego/superego, that corresponds to our contemporary understanding of neuroanatomy. The general conclusion we are led to by all of the research on Einstein's brain in the context of contemporary neuroscience is that we may never be able to demonstrate let alone prove a causal connection between Einstein's "genius" and the neuroscience of his brain.

Lepore is clearly aware of studies arguing for a non-neuroist approach in brain studies. He cites Steinmetz et al. (1994) study of twin pairs which demonstrated that "identical" (this term should not be taken literally; Restivo 2017: 84–86) genomes are not associated with identical neuroanatomies. This is an argument for explaining the development of the convolutions in the brain in non-genetic terms, that is, in terms of environment, experience, and chance. The neuroist response here is to underscore genomics in human biology and the mapping of the entire sequence of human DNA. In every case, the neuroist confronted with evidence for social, cultural, or environmental influences on the brain is never tempted to explore this evidence further. Instead, the tendency in such cases is to just turn to reinforcing some version of the brain-in-a-vat argument. It is important to keep in mind in this review of Einstein brain studies that it is diseased brains that have most profoundly shaped shared views among neuroscientists about the structure and function of the brain.

Intimations of the social are very fragile in the worlds of Einstein and his admirers. The classic Platonic notion of a pure mind is invoked in the image of Einstein needing only a piece of chalk and a blackboard to generate revolutionary thoughts. The materials on his family background, play things, friends, colleagues, and books are already a great concession to nurture. In order to see the significance of this, to have it stand out as a "factor," one has to be able to understand that the "lone thinker" is in fact a social system, a network of social relationships and cultural resources that provide the fuel for thinking alone. But the Platonic myth of a pure mind is reinforced by Lepore's focus on Einstein's reliance on the gedanken experiment. He gets even more reinforcement by valuing Einstein's own individualistic concept of mind and the "unknowable" origins of his creativity.

Neuroism, dissocism, and the myth of individualism blind investigators to the details of history. For this reason, Lepore, who has the additional disadvantage of being victimized by the myth of Kuhn's ideas on scientific revolution, can describe general relativity as a discovery sui generis.

Thought experiments have become increasingly influential in our era of "unverifiable science." Lepore cites string theory as an exemplar. What is often missed in discussions of thought experiments is that they are only useful if they are grounded in empirical and theoretically sound science and offer the possibility of being checked experimentally. Otherwise, they are simply ungrounded speculation with no potential for present or future utility in science. The value of Einstein's thought experiments was a consequence of his "feeling for the physical."

A key feature of Einstein's thinking about relativity theory is the assumption of a single observer possessing a single mind. This assumption simplifies the thought experiment. Lepore argues that the insights of relativity theory should be applied to the intra-observer/intra-cranial propagation of action potentials traversing the visual pathways. Thus the simple thought experiment which assumes single observer/single mind introduces unforeseen and unforeseeable inaccuracies for not acknowledging neuroanatomy/neurophysiology as the most likely basis of mind. Lepore makes two mistakes here. First, he assumes inaccuracies accrue to the simple thought experiment. But this is the condition of any experiment; we ignore in order to control and with the recognition that we are leaving out factors we do not know about that contribute to outcomes. The second mistake he makes is to adopt the neuroist assumption that mind must be rooted in neuroanatomy and neurophysiology. Again we find an intimation of the social in Lepore's unexamined assumption that there were "influences" impacting Einstein's creative achievements, presumably external ones. But the neurologist is stuck with the brain/mind equality assumption which leaves us with the so-called hard problem of how subjective experience emerges from a physical process. The solution is that it doesn't; it emerges from a social process which, like physical, chemical, and biological processes is a materialistic process.

Neuroism also supports the doctrine of cognitive individualism. The sheer quantity of neurons and the ongoing neuroplasticity of their myriad connections suggest that every individual thinks differently from everyone else. This ignores the demonstrable patterns of thought that characterize members of a given culture, network, or group. Lepore points out one of the noticeable features of Einstein's thinking; while he engaged in "pure thought experiments," he had a feeling for the physical that guided and grounded his thinking. One is reminded here that great scientists seem to have a feeling for their subject matter; notable examples are McClintock's "feeling for the organism," Tolman's "feeling for the rat," and Durkheim's "feeling for the social." Einstein's cultural capital included a great deal of work in physical labs, notably at the Zurich Polytechnic. His experience of wordless thought speaks to the levels of mentality that separate the wordless origins of thoughts and their materialization into vocalizable thoughts.

Conclusion

Many experts who studied Harvey's slides found nothing unusual about Einstein's brain. One of the reasons seems to have something to do with the particular way in which Harvey stained the tissues. It didn't allow for reliably distinguishing varieties of glia cells (astrocytes, oligodendrocytes, and microglia). Studies of myelinogenesis in the pre-connectome era were not on the neuroanatomists' radar. Harvey and his experts gave short shrift to myelin. No investigator in the past 60 years has taken on the daunting task of scrutinizing the microanatomy of Einstein's myelinated axons or dendrites. Counting the cells of Einstein's thalamus, a task offered to Sydney Schulman at Chicago, was considered too time consuming and not liable to offer any informative results. There was no significant research on Einstein's brain between 1955 and 1985.

The studies of Einstein's brain are at the end of the day sterile. They are guided by neuroist assumptions, a dissocist perspective, the fallacy of introspective transparency, and a conspiracy of mythologies: the myth of individualism, the myth of the brain in a vat; and the myth of brain-centric thinking reinforced by gene-centric thinking.

Culturally, we are accustomed to thinking about and experiencing "thinking" as a process that takes place inside of us and specifically inside our brains. I do not want to argue that we do not experience thinking in this way; we do. I do want to argue that this experience is misleading. The ethnography of shop talk in science, for example, demonstrates that the locus of thinking is talk among the scientists (Aman and Knorr-Cetina 1989). Collins (1998) has shown more generally that the locus of ideas is the social network. Thinking as an "outside" or "in-between" process does require a material site for its expression and this site is the brain/body. We are, as Gumplowicz (1980) recognized, voice boxes for expressing the thoughts generated in our social networks. Crane (2000) has put this in terms of thoughts arising at the nexus of biology and culture. Sociology allows us to ground the speculations of philosophers and psychologists of mind like Daniel Wegner and Peter Carruthers on the illusion of consciousness and free will.

Appendix: Case Study—Kekulé and the Dream that Stood on the Shoulders of a Social Network

Russell D. Larsen (1993), an epidemiologist, has written one of the most concise statements on how the sociology of science can reveal the social grounds of an episode of creative genius. His contribution to Wotiz's (1993) edited volume on *The Kekulé Riddle: A Challenge for Chemists and Psychologists* is aptly titled "Kekulé's Benzolfest Speech: A Fertile Resource for the Sociology of Science." Larsen mobilizes all of the key concepts in Robert K. Merton's sociology of science relevant to understanding the sociology of creative achievement. We begin with a classical tale of individual insight that seems invulnerable to sociological analysis: Kekulé's dream revealing the cyclic structure of benzene. Larsen constructs a new narrative of this event at the nexus of Merton's sociology of science and Kekulé's Benzolfest speech, "On Beyond the Ouroboros," delivered at the 1890 Berlin Benzolfest honoring the 25th anniversary of Kekulé's publication of the structure of benzene. In this speech, he revealed that he had discovered the ring shape of the benzene molecule after having a day-dream of a snake seizing its own tail (an ancient symbol known as the Ouroboros).

Let's begin by listing the Merton principles Larsen mobilizes:

- The Giants Aphorism (OTSOG: On The Shoulders of Giants): Merton (1965) traces this to Bernard of Chartres but earlier instances are lost in the mists of history. It is most famously stated by Isaac Newton: "If I have seen further, it is by standing on the shoulders of giants."
- Principle of Multiple Discovery: Inventions and discoveries are in principle multiples; singletons are rare.
- The Kindle-Cole Principle: Public controversy leads to battles for status; private controversy promotes the search for truth. The principle gets its name from the Hooke-Newton correspondence on their priority dispute. Hooke writes to Newton that if they take their dispute public it would "kindle cole (coal)" rather than uncover the truth of the matter.
- Anatopism: Failure to cite or otherwise refer to the original source of an idea on the assumption that the source is so well known that a citation or reference is unnecessary. The result is that the writer is credited with the idea.
- Larsen does not mention or use Merton's treatment of cryptomnesia[1]: On "cryptomnesia" see Merton (1973: 402–412).

- Matthew Effect: Crediting someone with an achievement she or he is not responsible for usually due to the halo effect or accumulated advantage (Merton cites Matthew's Gospel: "For to every one who has will more be given, and he will have abundance; but from him who has not, even what he has will be taken away;" Matthew 25: 29, *RSV*).
- Asparagus Effect (Ulam 1976): Trying to obtain more or all of the share of a joint effort for oneself.
- Heroic Theory or Great Man Theory: Explaining great achievements by labeling the creator a "genius." Credit accrues to the individual. In this book and elsewhere the heroic theory is opposed by social theory.
- Nationalism: The Newton-Leibniz priority dispute pitted the English establishment (Newton and his supporters) against the Hanoverians (Leibniz and the Leibnizian court philosophers).
- Priority Disputes: These are not rooted in humanitarian or personality considerations but rather arise from the institutional norms and structures of science that incline scientists to assert claims for their work.

How can we put these concepts to work in the case of Kekulé's dream? The dream narrative begins with Kekulé's remarks in his Benzolfest speech. He promises his audience "highly indiscreet disclosures from my inner life" that gave him the structure of benzene. The reasons we should doubt the dream narrative are given in Wotiz and Rudofsky (1993). If the dream narrative is a myth, what role does it play in the history of the discovery of the structure of the benzene molecule? Larsen claims that the dream narrative helped Kekulé establish the priority of his discovery in the context of a multiple discovery episode. Kekulé was not in fact the first one to propose the cyclic structure of benzene. His predecessors include J.J. Loseschmidt and A.S. Couper, among others. Kekulé fails to acknowledge directly the priority claimants but invokes the "giants aphorism" and the principle of multiples. This allows him to acknowledge that "certain ideas at certain times are in the air"; that his ideas "came from seeds that had previously been sown" and that there is "no such thing as novelty in the matter." The "giants aphorism" allows Kekulé to indirectly acknowledge predecessors (the giants) but to give himself credit for standing on their shoulders and seeing further (establishing his priority). He liberally cites contributors to the benzene question but not the key contributors to the discovery of the structure of benzene. Larsen claims that Kekulé was known for a tendency to selectively acknowledge others and in the benzene case he may have been anatopic. There is further evidence of Kekulé's misconduct in pursuit of his priority claims (Wotiz and Rudofsky 1993: 262–264).

With respect to the Matthew Effect, it appears that Kekulé got credit for the discovery because he had greater visibility than other potential claimants. Furthermore, Kekulé refers to the role of genius in discovery without specifically claiming to be a genius. Larsen sees this as a case of "thinly-veiled self-aggrandizement" especially given the fact that it is delivered in the Benzolfest speech.

The Kindle-Cole principle is evoked in the decision to pursue the debate between Kekulé and Ramsay and Rocke over the dream narrative through private correspondence instead of a public debate.

Nationalistic zeal is reflected in the conflict between the German nationalist Herman Kolbe and Kekulé who Kolbe accused of being an internationalist. The personal animosity showed up in Kolbe's editorials for the *Journal Für Praktische Chemie*. Kolbe attacked Kekulé's politics and also penned derogatory remarks about his dream narrative using terms like fantasy, figments of imagination, and deceptive speculations.

There is some evidence that Kekulé knew of and used Loseschmidt's work and to the extent that this is true and is part of the multiple discovery context we have an example of the asparagus effect. Ulam (1976) named this effect based on a boardinghouse story of a man who ate most of a serving of asparagus. Another man pointed out that there were others at the table who also liked asparagus.

Merton also discussed adumbrationism.[2] This shows up in Kekulé's remark in the Benzolfest speech that the explorer will never find anything essentially new. Kekulé's dream narrative is a portal to a number of sociological insights into the nature of science. The point is not to discredit Kekulé but to ground his achievements by contextualizing them in a sociologically realistic framework.

Bibliographic Notes for Chapter 3

- Documentaries and Websites
  On Einstein's brain as a relic, see Hull (1994); see Kremer (2015) on the strange afterlife of Einstein's brain, and the NPR (National Public Radio) Morning Edition program (2005); also see Levy

(1978) and Edmonds (2008). The story of Ishi, "the last of the Mohicans" (in fact, the last [?] member of a tribe wiped out in the California genocide), is told in many books and films: see Kroeber (1961), and Fleras (2006). Theodora Kroeber was the wife of Alfred Kroeber, one of the prominent anthropologists who studied Ishi over a period of years. There are numerous ethical and scandalous issues raised by this episode in the history of anthropology (see references cited above).

- Selected References
  Abraham (2002); Balter (2012); Diamond, Scheibel, Murphy, Jr., and Harvey (1985); Falk, Lepore, and Noe (2013); Hotz (2005); Men, Falk, Sun et al. (2014);
  And Witelson, Kigar, and Harvey (1999). NOTE: For an exhaustive list of references to studies of Einstein's brain, see Lepore (2018); and see Paterniti (2000). On Kuhn's widely admired but misunderstood theory of scientific revolutions, see Kuhn 1962, 1970; and Restivo's critique (1983).
- On the breakdown of the body/brain/mind divisions, see Brothers 1997, 2001; Pert 1997; and Rose and Rose 2016.
- Environmental Enrichment/Enhancement
  Halperin and Healey (2011); Pang, Terence & Hannan, Anthony (2012); Mohammed, Zhu, Darmopil et al. (2002: 109) and other contributions to Hoffman, Boer, et al. (2002). In spite of the predominance of brains-in-a-vat and neuroist paradigms, the idea that external stimuli can cause structural and functional changes in the brain was already considered by the ancient Greeks. This has been observed in two related fish, one that inhabits caves the other swamps and streams, and in domesticated and wild rats, turkeys, pigs, and silver foxes. Against this background came the unexpected finding that small changes in the housing environments of adult laboratory rats could induce changes in brain chemistry and anatomy. And see Carey (2010) on genes as mirrors of life experiences: "By studying genes at the 'epi' level, scientists are hoping to discover patterns that have been elusive at the level of the genes—and ideally to find targets for calibrated treatments that would not simply shut off errant genes but would gradually turn their activity up or down, like adjusting the balance on a stereo." Two of the most important contributions to my thinking about genes and genetics are David Moore (2002, 2015) and Shea (2008).

Notes

1. The word *cryptomnesia* is a portmanteau of *crypto-* (which comes from the Greek word *kryptos* meaning "hidden, concealed, secret") and *amnesia*. The word was first used to explain latent memories in psychics. Psychiatrists Jung, Dukes, Ferenczi, and Freud treated the subject, and there have been empirical studies of the phenomenon. There are two types of cryptomnesia, coming under the general concept of plagiarism: (1) The plagiarizer regenerates an idea that was presented earlier, but believes the idea to be an original creation; and (2) the ideas of others are remembered as one's own.
2. Merton distinguishes "adumbration," foreshadowing elements in new research and "adumbrationism," the ill-intentioned and sometimes occult search for ancient anticipations of new research. On the concept of adumbrationism, see Merton (1961: 474). The concept is also discussed in Sztompka (1991); see the extended discussion in Merton (1968: 13–27). For an historical case study of adumbrationism in the history of science and religion, see (Restivo 2018: 150–173).

References

Abraham, C. 2002. *Possessing Genius: The Bizarre Odyssey of Einstein's Brain*. New York: Vintage.

Aaman, K., and Karin Knorr-Cetina. 1989. Thinking Through Talk: An Ethnographic Study of a Molecular Biology Laboratory. *Knowledge and Society* 8: 3–26.

Balter, M. 2012. Rare Photos Show that Einstein's Brain Has Unusual Features. *The Washington Post*, Tuesday, 27 November E6.

Bertolero, M., and D. Bassett 2019. How Matter Becomes Mind. *Scientific American* July: 26–33.

Brothers, L. 1997. *Friday's Footprint: How Society Shapes the Human Mind*. New York: Oxford University Press.

———. 2001. *Mistaken Identity: The Mind-Brain Problem Reconsidered*. Albany, NY: SUNY Press.

Carey, B. 2010. Genes as Mirrors of Life Experiences. *The New York Times*, November 8. https://www.nytims.com/2010/11/09/health/09brain.html.

Clark, A. 1998. *Being There: Putting Brain, Body, and World Together Again*. Cambridge, MA: Bradford.

Collins, R. 1992. *Sociological Insight: An Introduction to Non-Obvious Sociology*. 2nd ed. Oxford: Oxford University Press.

———. 1998. *The Sociology of Philosophies*. Cambridge, MA: Harvard University Press.

Colombo, J., et al. 2006. Cerebral Cortex Astroglia and the Brain of a Genius: A Propos of A. Einstein's. *Brain Research Reviews* 52 (2): 257–263.

Connolly, J.C. 1950. *External Morphology of the Primate Brain*. Springfield, IL: C. C. Thomas.

Crane, M.T. 2000. *Shakespeare's Brain*. Princeton: Princeton University Press.

Damasio, A. 1994. *Descartes' Error*. New York: G.P. Putnam.

Diamond, M.C., A.B. Scheibel, G.M. Murphy, and T. Harvey. 1985. On the Brain of a Scientist: Albert Einstein. *Experimental Neurology* 88: 198–204.

Edmonds, M. 2008. How Albert Einstein's Brain Worked. *HowStuffWorks.com*, October 27. https://science.howstuffworks.com/life/inside-the-mind/human-brain/einsteins-brain.htm.

Falk, D., F.E. Lepore, and A. Noe. 2013. The Cerebral Cortex of Albert Einstein: A Description and Preliminary Analysis of Unpublished Photographs. *Brain: A Journal of Neurology* 136 (4): 1304–1327.

Fleras, A. 2006. Ishi in Two Worlds: A Biography of the Last Wild Indian in North America. *Journal of Multilingual and Multicultural Development* 27 (3): 265–268.

Franklin, S. 1995. *Artificial Life*. Cambridge, MA: MIT Press.

Galaburda, A. 1999. Albert Einstein's Brain. *Lancet* 354 (9192): 1821–1823.

Geertz, C. 1973. *The Interpretation of Cultures*. New York: Basic Books.

Golden, F. 1999. Albert Einstein: Person of the Century. *Time* 154 (27): 62–65.

Gumplowicz, L. 1980. *The Outlines of Sociology*. New Brunswick, NJ: Transaction Books, 1980; orig. publ. in German, 1885.

Halperin, J.M., and D.M. Healey. 2011. The Influences of Environmental Enrichment, Cognitive Enhancement, and Physical Exercise on Brain Development: Can we Alter the Developmental Trajectory of ADHD? *Neuroscience & Biobehavioral Reviews* 35: 621–634.

Herculano-Houzel, S. 2014. The Glia/Neuron Ratio: How It Varies Uniformly Across Brain Structures and Species and What That Means for Brain Physiology and Evolution. Wiley Online Library (wileyonlinelibrary.com). https://doi.org/10.1002/GLIA.22683.

Hines, T.H. 2014. Neuromythology of Einstein's Brain. *Brain Cognition* 88 (July): 21–25.

Kandel, E.R., J.H. Schwartz, et al., eds. 2013. *Principles of Neural Science*. 5th ed. New York: McGraw-Hill.

Kantha, S.S. 1992. Albert Einstein's Dyslexia and the Significance of Brodmann Area 39 of His Left Cerebral Cortex. *Medical Hypotheses* 37 (2): 119–122.

Kroeber, T. 1961. *Ishi in Two Worlds*. Berkeley: University of California Press.

Kuhn, T.S. 1962. *The Structure of Scientific Revolutions*. Chicago: University of Chicago Press, 2nd ed., 1970.

Larsen, R.D. 1993. Kekulé's Benzolfest Speech: A Fertile Resource for the Sociology of Science. In *The Kekulé Riddle: A Challenge for Chemists and Psychologists*, ed. J.H. Wotiz, 177–193. Vienna, IL: Cache River Press.

Lepore, F.E. 2018. *Finding Einstein's Brain*. New Brunswick, NJ: Rutgers University Press.

Levy, S. 1978. I Found Einstein's Brain. *New Jersey Monthly*, August. https://njmonthly.com/articles/historic-jersey/the-search-for-einsteins-brain/.

Men, W., D. Falk, T. Sun, W. Chen, J. Li, D. Yin, L. Zang, and M. Fan. 2013. The Corpus Callosum of Albert Einstein's Brain: Another Clue to his High Intelligence? *Brain: A Journal of Neurology* 137 (4): e268. https://doi.org/10.1093/brain/awt252.

Merton, R.K. 1961. Singletons and Multiples in Scientific Discovery: A Chapter in the Sociology of Science. *Proceedings of the American Philosophical Society*, 105: 470–486; reprinted in R.K. Merton (1973). *The Sociology of Science*. Chicago: University of Chicago Press, 1973, pp. 343–370.

———. 1965. *Standing on the Shoulders of Giants: A Shandean Postscript*. New York: The Free Press.

———. 1968. *Social Theory and Social Structure*. New York: The Free Press.

———. 1973. *The Sociology of Science*. Chicago: University of Chicago Press.

Moore, D. 2002. *The Dependent Gene: The Fallacy of "Nature vs. Nurture"*. New York: Henry Holt and Company.

———. 2015. *The Developing Genome: An Introduction to Behavioral Epigenetics*. Oxford: Oxford University Press.

Noé, A. 2010. *Out of Our Heads: Why You Are Not Your Brain, and Other Lessons from the Biology of Consciousness*. New York: Hill & Wang.

Paterniti, M. 2000. *Driving Mr. Albert: A Trip Across America with Einstein's Brain*. New York: Dial Press.

Putnam, H. 1981. *Reason, Truth, and History*. Cambridge: Cambridge University Press.

Restivo, S. 1983. The Myth of the Kuhnian Revolution in the Sociology of Science. In *Sociological Theory*, ed. R. Collins, 293–305. New York: Jossey-Bass.

———. 2017. *Sociology, Science, and the End of Philosophy: How Society Shapes Brains, Gods, Maths, and Logics*. New York: Palgrave Macmillan.

———. 2018. *The Age of the Social: The Discovery of Society and the Ascendance of a New Episteme*. New York: Routledge.

Restivo, S. 2005/2019. Romancing the Robots: Social Robots and Society, pp. 10, 17. Available at salrestivo.org.

Rose, S., and H. Rose. 2016. *Can Neuroscience Change Our Minds?* Cambridge: Polity Press.

Searle, J. 1992. *The Rediscovery of Mind*. Cambridge, MA: MIT Press.

Seung, S. 2012. *Connectome: How the Brain's Wiring Makes Us Who We Are*. New York: Houghton Mifflin Harcourt.

Shea, E.P. 2008. *How the Gene Got Its Groove*. Albany, NY: SUNY Press.

Starns, O. 1994. *Ishi's Brain*. New York: W.W. Norton.

Steinmetz, H., A. Herzog, et al. 1994. Discordant Brain-Surface Anatomy in Monozygotic Twins. *New England Journal of Medicine* 331: 952–953.

Sztompka, P. 1991. *Society in Action: The Theory of Social Becoming*. Chicago: University of Chicago Press.

Swaab, D.F. 2014. *We Are Our Brains: A Neurobiography of the Brain, From the Womb to Alzheimer's*. New York: Spiegel & Grau.

Thompson, E., and D. Cosmelli. 2011. Brain in a Vat or Body in a World? Brainbound Versus Enactive Views of Experience. *Philosophical Topics* 39 (1): 163–180.

Ulam, S.M. 1976. *Adventures of a Mathematician*. New York: Charles Scribner's Sons.

Williams, D.L. 2002. *The Mind in the Cave*. High Holborn, UK: Thames and Hudson.

Witelson, S.F., D.L. Kigar, and T. Harvey. 1999. The Exceptional Brain of Albert Einstein. *The Lancet* 353 (9170): 2149–2153.

Wotiz, John H., and Susanna Rudofsky. 1993. Herr Professor Doktor Kekulé: Why Dreams? In *The Kekulé Riddle: A Challenge for Chemists and Psychologists*, ed. J.H. Wotiz, 247–276. Vienna, IL: Cache River Press.

Zaidel, D. 2001. Neuron Soma Size in the Left and Right Hippocampus of a Genius. [Conference Poster: http://cogprints.org/1927/].

CHAPTER 4

Genius: Standing on the Shoulders of Social Networks

Abstract This chapter critically situates the idea of "genius." The very idea of genius is based on and reinforces the myth of individualism and the "I" as a grammatical illusion. As a sociologist, I claim that if you give me a genius, I will give you a social network. I illustrate this claim with brief looks into non-Euclidean geometry, Ramanujan, Nikola Tesla, and Rodin followed by an exploration of the Einstein genius cluster. Other topics include chaos and creativity, the social context of genius, intuition, and an appendix on creativity and madness. I examine the key notions of genius clusters and the multiples idea: any given innovative idea or technology appears along with more or less similar innovations—families of innovations or parts thereof—at the same time within the boundaries of a cultural or civilizational region.

Keywords Genius • Social network • Chaos • Intuition • Madness

A Nietzschean Overture

The view of "genius" that guides my understanding of Einstein and his brain is grounded in a sociological concept of what it means to be human. And there is the nub of it: the genius is "human, all too human." In Nietzsche's (1996/1878–1880) work with this title we find the basic ingredients of a sociology of genius. Everything we humans do and think, even if we are labeled "genius," is incredibly complicated but none of it is

S. Restivo, *Einstein's Brain*,
https://doi.org/10.1007/978-3-030-32918-1_4

miraculous. It would be beneficial for "geniuses" themselves to grasp the human qualities and fortunate circumstances that have come together in them. At the end of the day, the "genius" as an individual is no more a miracle than is any serious workman. There are dangers to the "genius" and to his admirers who still believe in magic and miracles in the view that the genius is superhuman. *Human, All Too Human* is seasoned with insights on "the cult of genius" and its grounding in the failure to attend to the "becoming" of the works of the "genius." There it is; Nietzsche is the locus classicus for the "I" as a grammatical illusion and the "genius" as human all too human, never mind free will as one of humankind's original errors: "Geniuses know better than the talented how to hide the barrel-organ [a mechanical musical instrument from which pre-programmed music is produced by turning a handle] by virtue of their more abundant drapery; but at bottom they too can do no more than repeat their same old tunes" (Nietzsche 1996/1878–1880: 248).

Introduction

My objective in this chapter is to critically situate the idea of "genius" or what appear to be mysterious and awe inspiring cases of individual creativity. The very idea of genius is based on the myth of individualism and the "I" as a grammatical illusion. Can we in fact sustain the common conception of Einstein as a lone wolf genius? My claim as a sociologist is that if you give me a genius I will give you a social network. The genius can be thought of as a node in a social network or to revise a classical metaphor as standing on the shoulders of a social network. I showed in Chap. 3 how my concept of the person in the social network is different from traditional ways of thinking about social networks.

That "genius" or creative achievement doesn't make sense as a lone wolf phenomenon is grounded in four fundamental ideas: (1) Genius clusters: geniuses do not appear at random and scattered unpredictably across time, space, and culture; (2) Genius clusters do not appear at random but during the waning periods of nations or civilizational areas or during periods of florescence (a renaissance, a blooming of a thousand flowers); (3) Innovative ideas and technologies appear as multiples not singletons: that is, any given innovation will appear along with more or less similar innovations—families of innovations or parts thereof—at the same time within the boundaries of a cultural or civilizational region; (4) Humans come onto the evolutionary stage already, always, and everywhere social:

they don't arrive as individuals who then become social but rather as social animals who then experience various degrees of individuation depending on eras and cultural configurations.

We should expect geniuses to be, like all humans, embedded in society not separate from it. As evidence for this idea, I introduce brief vignettes of several cases in which ab novo innovations turn out to have histories and innovators are shown to stand on the shoulders of social networks. My approach will not be exhaustive. My objective is to demonstrate that a network with precursors and collaborators exists. The cases I consider are the invention of non-Euclidean geometries; the Indian mathematician Ramanujan; the "master of lightning," Nikola Tesla; the sculptor Auguste Rodin; and Einstein, "the genius of geniuses."

Part I: Give Me a Genius and I Will Show You a Social Network

Non-Euclidean Geometry

Let's begin with non-Euclidean geometry which in fact plays a key role in the development of relativity theory. Mathematics has been shrouded in mystery and divinity throughout its history. Mathematics is probably populated with as many or more geniuses in the popular imagination than physics, the arts, and music. The mystery around mathematics has easily slipped into the worlds of magic, witchcraft, and the occult and the halo atop mathematicians has sometimes transformed into the horns of the devil. In his "Deliverance from Error," the Arab scholar Al-Ghazali (1058–1111) reflecting in part the political circumstances of his time (the Caliphates were absorbed in wars and ignored science and technology) pointed out two bad results of mathematics. Students of mathematics, Al-Ghazali warned, were in danger of assuming that the clarity and solidity of mathematical proofs could be generalized to all departments of philosophy. If mathematics was capable of honing the keenest intellects, it would have led to discerning the truth in religion. Then there is the danger that when the everyday person hears that these learned mathematicians with their proof methods have rejected religion he or she will be inclined to reject religion. Then he or she may encounter sincere but ignorant Muslims who defend religion by rejecting science. When the mathematicians and philosophers hear this, they become stronger in their resistance to religion and stronger in their defense of their proof-based views.

Science watchers along with scientists and the educated among us are all familiar with Galileo famously saying "The book of nature is written in the language of mathematics." Many modern scientists agree with Galileo. Kant rejected as science any field that was not based on mathematics. John Dee (1527–ca. 1608) lived too early to know anything of Galileo but would likely have agreed with his view of mathematics as the language of nature. The problem is that Dee was variously considered by his contemporaries as a philosopher, an astrologer, a magician, and above all a mathematician. This was not an unusual image of the proto-scientist natural philosopher in the centuries leading up to the scientific revolution and even afterwards. There were no mathematicians per se but rather number workers who were at once numerologists, astrologers, and alchemists. Dee as a number worker embodies the dark side of the math coin for he used it for horoscopes, numerology, alchemy, and occult exercises designed to communicate with the angels. He was nonetheless a strong defender of mathematics in a Tudor world that was suspicious of mathematics, and considered it disreputable and allied with witchcraft. Copernicus aroused suspicion not simply for his heliocentric proposal but his use of mathematics to deduce what could not be directly seen. Mathematics books were often viewed as "conjuring books." The mystery, purity, and transcendence of mathematics have been nourished from its earliest beginnings in the works of Pythagoreans and Platonists. Its practitioners still carry some of the qualities of the magician. Numbers can be "invisible," "imaginary," "irrational," and hidden in "magic squares."

The manipulation of number facts in "mysterious" ways for audiences who are not literate in math is part of the toolkit of many stage magicians. So it shouldn't be entirely surprising that the dean of twentieth-century historians of mathematics Carl Boyer and the late Dirk Struik, Marxist, mathematician, historian of mathematics, and MIT professor both could describe the emergence of non-Euclidean geometries (NEGs) as a miracle of individual genius. Their narrative is the story of three individuals, Lobachevsky, Riemann, and J. Bolyai, isolated from each other and outside the centers of late nineteenth-century mathematics, independently and more or less suddenly inventing NEGs ab novo. So Boyer and Struik give me geniuses. I said, in such a case I would give them a social network.

The historical sociology of NEGs tells a story that is quite different from the one told by Boyer and Struik (notable among others). The network I give you has Gauss and the University of Göttingen at its center. Lobachevsky was indeed on the outskirts of the mathematical community at the University of Kazan. But that university's mathematicians included Gauss' teacher J.M. Bartels and other distinguished German mathematicians. Riemann did his dissertation under the direction of Gauss. And J. Bolyai was the son of W. Bolyai, one of Gauss' friends and a colleague at Göttingen. This is the basic outline of the social network that gave rise to NEGs.

The significance of the Gauss-Göttingen connection is that Gauss had been working on NEGs since the late 1700s. He corresponded with other mathematicians about NEGs throughout the early 1800s and published notes on the topic from 1831 on. This is the late phase of a history of NEGs that begins with Euclid's earliest commentators on the parallels postulate and unfolds more or less continuously in a history marked by such names as Saccheri (1667–1733) and Lambert (1728–1777). What looks on the surface and through the lens of the cult of the individual like isolated genius turns out on closer examination to be a social network.

Ramanujan

Ramanujan (1887–1920) is another apparent isolated genius, an autodidact who constructed numerous valid mathematical theorems without any formal training in or exposure to mathematics. The anti-narrative begins when we learn that he had two lodgers during his pre-teen years who were mathematics students. Furthermore, he was able to borrow a copy of S.L. Loney's text on advanced trigonometry. At the age of 16, he came across a library copy of G.S. Carr's *A Synopsis of Elementary Results in Pure and Applied Mathematics*. This book was a collection of 5000 theorems. Once again, the most cursory examination of the life of this "isolated genius" turns up connections to people and books relevant to his creative acts. My conjecture is that the key ingredient in this admittedly limited network was Carr's book.

I conjecture that Ramanujan, through persistent engagement with the theorems in this book recognized certain patterns, and was able to construct a theorem generating algorithm. This algorithm led to the construction of his theorems, both those that proved to be true and those that proved in the end to be false. This is better explained in any case by persistent, focused attention and hours upon hours of study than by simply labeling Ramanujan a "genius."

During a hospital stay while in England Ramanujan was visited by his Cambridge mentor and friend G.H. Hardy. The visit led to the famous

1729 episode. Hardy referred to the number on his taxi, 1729, as a number with no notable features. Ramanujan immediately pointed out that on the contrary 1729 is the smallest number that can be expressed as the sum of two cubes in two different ways. Ramanujan was in fact already familiar with this number and its properties but the story tends to be told in a way that emphasizes insightful genius rather than hard work. None of this in any way diminishes Ramanujan's significant achievements. Those achievements stand, like Einstein's, whether we ground them in the myth of genius or the reality of hard work and social networks.

Evelyn Fox Keller popularized the phrase "a feeling for the organism" in her book on the Nobelist cytogeneticist Barbara McClintock. It is clear that the great scientists have a "feeling for" their subject matter. In one sense, the association of "feeling" with great science hints at something sexist or at the very least gendered. But a "feeling for" the scientist's subject is characteristic of great scientists, men and women alike. McClintock had a feeling for the maize; Einstein had a feeling for the photon and for "the physical"; Edward Tolman had a feeling for the rat faced with a maze; Ramanujan had a feeling for the number; and Tesla had a feeling for electrical phenomena.

Nikola Tesla

Nikola Tesla (1856–1943) has a claim to Einstein's title as "genius of geniuses." He is certainly in any case a classic case of the genius. He has been called a wizard, the master of lightning, and the inventor of the twentieth century. There are intimations in this sort of description of a connection between genius and the occult. I was drawn to his work during the years I was an electrical engineering student at Brooklyn Technical High School and the City College of New York. I was especially intrigued by the prospects for wireless transmission of power and energy reported in his notes and papers.

I want to consider just one aspect of Tesla's wide-ranging contributions, alternating current. Tesla's AC network is tied to the institution of the patent. Just as the energy-mass equation had a pre-Einsteinian history, so did alternating current have a history before Tesla. The first generation of electromagnetic generators was invented in the 1820s. Their rotating mechanical motors generated alternating current. All the electrical devices in this period ran on direct current. The AC machines were thus fitted with commutators that transformed AC to DC. AC was already widely used across the United States several years before Tesla patented his AC system. Lucien Gaulard and John Gibbs designed transformers for Westinghouse and by 1885 the company was selling AC power systems designed by William Stanley and others. The AC-DC "War of the Currents" pitting Tesla (AC) against Edison (DC) was already on before Tesla even designed his system. Edison was already raising objections to AC by 1886. He initiated a series of public demonstrations of the dangers of AC by hiring Harold Brown to design an electric chair.

Tesla was not alone in inventing the twentieth century. A Tesla genius cluster for the middle to late nineteenth and early twentieth centuries would include besides Edison and Westinghouse, Luther Burbank (new plants varieties), Joseph Gayetty (toilet paper), John Froehlich (gas powered tractor), Alexander Graham Bell (telephone), Samuel Morse (telegraph and Morse code), Alfred Nobel (dynamite), Humphrey Davy (first electric light), and Louis Pasteur (pasteurization). And this cluster ties in with the Einstein cluster and the 1840–1920 "genius cluster." This brief vignette puts us on the road to constructing the social networks whose shoulders Tesla stood on.

Auguste Rodin

What do you see when you see Rodin's famous sculpture, *The Thinker* (*Le Penseur*)? Do you see a great piece of art created by Rodin? Perhaps that's not the end of the story. Consider the film, Camille Claudel. Claudel was Rodin's student and lover but above all a great artist herself. In one section of the film, Claudel is shown entering the studio where *The Thinker*, as one of the elements in the sculpture garden, *The Gates of Hell*, is being sculpted. The question is: Where is Rodin and what is he doing? As Claudel walks into and through the studio, we see many people climbing up and down ladders, standing on scaffolds and on the ground chipping away at chunks of marble and other materials. We see materials being brought into the studio and taken out. There is, in brief, a great deal of collective activity taking place and we see this all along Claudel's walk through the studio. Again, where is Rodin and what is he doing? We finally come across him in the midst of all this hubbub of activity wearing a suit and tie and a hat talking to representatives of the Ministry of Finance. What does the clip show?

First let's understand that the film is an artistic account and not a documentary. At this point at least it is virtually an ethnography of Rodin's studio. What we see is that if we look at *The Thinker* and only see the work of one man, Rodin, we will fail to understand what we are looking at. We have to see *The Thinker* as embodying all of the activity in that studio directed toward construction of the sculpture along with all the other sculptural ingredients of *The Gates of Hell*. Rodin deserves credit not for being the sole creator of *The Thinker* but for being the key node in a social network of workers all of whom have stamped their activities into and on the sculpture(s). And remember that even when Rodin was working on the sculpture "alone," he was working as a "self as social structure" and not as an instantiation of the grammatical "I."

The "But What About" Retort

Stories similar to the ones I have told here can be told about every "genius," every "lone," "isolated" innovator and about every idea, person, or thing whether a tea pot, an equation, the Empire State Building, the Queen Mary, the Mona Lisa, Einstein, you, or a handkerchief; They are all manufactured in social networks. I have not told the full story of the cases I review here. I have simply opened the door enough to make it clear that networks abound on the other side of the "lone wolf" door. Of course, even if I'd spun more detailed narratives, someone would always be able to pose the "but what about" question. Sure, I've convinced you that there's something to the social network idea in the cases I've reviewed but "what about X?" The implication of the cases I've reviewed is that they offer a rationale for the mantra, "give me a genius and I will show you a social network." The fact that I haven't examined each and every case of "genius" means that there is a reservoir of "but what about" cases on which to test the hypotheses or conjectures about genius clusters, multiples, and social networks. I've already established that the "I" is a grammatical illusion. That means that in every "but what about" case, we are dealing already with individuated social beings. Even when I am standing alone painting a chapel's ceiling in a church, I am a social network. Like Einstein, Michelangelo's creative acts are those of a self that is a social structure. It is Michelangelo the socially structured self that paints when he is alone, not Michelangelo the lone wolf individual.

Part II: Albert Einstein—Genius of Geniuses

What is at stake when we define Einstein as a "genius?" What does this actually tell us about Einstein? First of all the term "genius" enhances and exaggerates the concept of the individual a- or non-social self. Second, it sets Einstein apart from the rest of us in an intellectual world we could never hope to understand let alone inhabit ourselves. For the concept of "genius" to be meaningful, for it to mean scientifically what it conveys to the general public in everyday terms, it would have to be rooted in genes, neurons, or both, or else a soul or some ineffable "inner light." In that case, geniuses would appear at random and scattered across the intellectual landscape. I pointed out earlier that geniuses do not appear at random. Genius clusters. If we absorb "genius" under the somewhat more easily operationalized concept of "creativity," the clustering of creative periods was already recognized by the Roman historian, Marcus Velleius Paterculus who died around 31 CE. Furthermore, periods of creative clusters appear predictably during periods of rapid decline or rapid growth within civilizations or cultural areas.

The findings on genius clusters converge with research in the sociology of science on inventions and discoveries. All inventions and discoveries are in principle, multiples, not singletons. Often the individual who gets credit for a multiple is simply someone in the right place at the right time. This helps to explain some of the criticisms of the uniqueness of Einstein's discoveries. The practice of eponymy illustrates an aspect of this process. The person for whom we name an idea, a law, or theorem is often not the person standing on the shoulders of a social network who deserves the credit if we are going to play the credit game. And indeed, as should be obvious by now, identifying a single individual with an innovation is a problematic exercise.

Einstein's 1905 papers come in the midst of a cultural flowering of ideas, inventions, and discoveries across the full spectrum of the arts, humanities, and sciences between 1840 and 1920. Einstein's "genius" cluster in physics, science, and technology included such luminaries as Planck, Tesla, Marconi, Westinghouse, Madame Curie, the Wright brothers, Emmy Noether, Lorentz, Poincaré, Minkowski, and Edison. The two great innovations in physics that would remain at the core of physics throughout the twentieth and into the twenty-first century, relativity theory and quantum mechanics, are born in the early years of the twentieth century. What is the social network of $E = MC^2$?

The story I grew up reading and hearing about as a school boy was that Einstein, isolated from the major centers of activities in physics as a patent clerk in a Swiss patent office, suddenly burst upon the world scene with his theory of relativity in 1905, shocking the worlds of physics, scientists and intellectuals, and the lay public. As an undergraduate and later in graduate school my readings in the history of physics introduced me to the works of the central figures in late nineteenth- and early-twentieth-century physics and mathematics. I learned that relativity theory had been a part of the history of physics long before Einstein came along. Indeed, if we define relativity theory broadly enough, its history goes back to the origins of the scientific revolution and beyond. There were so many precursors to Einstein's work that the charge of plagiarism emerged in the minds of some physics watchers and Poincaré himself.

What is the relativity theory network? Forms of relativity theory can be traced back to da Vinci, Galileo, Giordano Bruno, and others. Let's consider in particular the iconic equation $E = MC^2$. Given the framework I've been developing, we must assume that that equation did not emerge full blown ab novo in the mind of Albert Einstein sometime around 1905. There must be a path that leads to Einstein as the locus of a specific configuration of social networks, "genius clusters," and a blossoming cultural context. In the context of those specifics that are in a real sense who Einstein is, Einstein did something novel. That novelty emerged from a specific pathway along which he accumulated a specific set of resources. Einstein, as the point of the convergence of these specifics, becomes the locus for the idea that there could be an energy-mass transfer linked by the conversion factor c. How was this point reached?

Let's begin with Newton and Leibniz. Newton had argued that the analysis of two objects making contact is given by the product of their mass times their velocity, mv^1. This was the position Voltaire defended in his popular renditions of Newton's works. The indomitable Madame du Châtelet, author of what is still considered the definitive French translation of Newton's Principia, took up Leibniz's view. Newton's mv^1 left certain gaps in the physics of colliding objects, gaps that meant God had to intervene from time to time to wind up the clockwork universe. Leibniz argued that the collisions should be analyzed using mv^2. This is the position du Châtelet defended. She was Voltaire's collaborator and mistress but more importantly a great natural philosopher in her own right whose contributions have been overshadowed by Voltaire's more potent image and personality. In Newton's equation energy disappears; thus the need for God's intervention. Nothing disappears using Leibniz's analysis.

The idea of squaring a number was an ancient idea. A 4X4 patch of tiles contains 16 tiles not 8. Leibniz's mv^2 puts us on the path to mc^2. First, du Châtelet was able to defend her opposition to Voltaire thanks to a Dutch study in which weights were allowed to fall into soft clay. The researchers found that if a small brass ball was sent down twice as fast as in the first trial it went four times as deep into the clay. Three times faster and it sank nine times as far. This offered support for the equation $E = mv^2$. These results reflected something about the geometry of our world which gives rise to squared numbers in many natural situations. We have the Dutch results; on a more mundane level, it you move twice as close to a reading lamp the light intensity increases four times. This turns out to be the case in most accumulations; they unfold according to the principle of squares. This is the basic feature of nature that eventually leads Einstein to settle on c^2 as the correct energy-mass conversion factor. The matter is more complicated and involves multiple pathways. For a more technical answer to this question, see Rathore (2018). Briefly, we square the speed to get the units right. Getting the units right is extremely important and often is enough to solve a problem. Why the speed of light? Because it is the only invariant in relativity theory (invariantentheorie).

In Einstein's time, Poincaré discussed the principle of relativity in papers presented in 1904 and 1905. In these papers, Poincaré had energy-mass equivalence, an equivalence already published by Olinto De Pretto in a 1903 paper (Bartocci 1999). Samuel Taylor Preston had speculated on mass-energy equivalence as early as 1875. We have in these examples the core of the multiples context within which Einstein worked. So now when we ask "Why Einstein and not someone else?" we have an answer. The answer is not that Einstein was genetically or neuronally gifted in such a way that only he could develop the special and general theories. The answer is that Einstein stood on the shoulders of a specific social network within a specific "genius cluster" grounded in a specific scientific and artistic cultural renaissance. Einstein was the sum of those specifics. Recall the details of the Einstein network outlined in Chap. 1.

Chaos and Creativity

The significance of genius clusters and multiples is that they generate things and ideas in a concentrated proliferation that allows for innovative interactions, combinations, and re-combinations, the basic foundations of innovation and in our times the engines of modernity. There is a certain chaos in this proliferation, and chaos seems to be part of the narrative of creative, innovative achievement. It appears then that there is a social chaos that is reflected in the chaos of the individual creator's narrative in life and in innovation. It is part of the innovator's narrative as an expression of the network's chaos, but it can also be cultivated and nourished by self-aware individuals. Neurologically, social chaos is coupled with chaotic brain states that process information in novel ways and re-order the brain for innovation. I say more about this in Chap. 5.

In the next section of this chapter I turn to the question, "Where are the women Einsteins?" We live in an era in which the awareness of sex and gender biases has spread across all segments, institutions, occupations, and professions of society. When Einstein's name comes up in the context of genius narratives, how many people will make the association to the mathematics genius, Emmy Noether (1882–1935)? Einstein, among others, described her as the most important woman in the history of mathematics. Noether's theorem connects symmetry and conservation laws. She is described as a "genius" in many of the online references to her and her work. Nonetheless, genius doesn't make much room for women and I want to say something about why the very idea of genius closes its doors to women.

Gaging Gender and Genius

> There is no female mind. The brain is not an organ of sex. As well speak of a female liver. Charlotte Perkins Gilman (1860–1935), Women and Economics (1898)

In order to illustrate and problematize gender and genius we can follow the case study of two philosophers. Christine Battersby argues that the genius can be all sorts of men but never a woman. Peter Kivy, on the other hand, argues that there are two types of genius rooted in the philosophies respectively of Plato and an anonymous first-century author (sometimes supposed to be Longinus), possessed and possessor. The Platonic genius is possessed in the sense I introduced earlier: possessed, inhabited, by a spirit, a god, a guardian angel. The Longinus genius is the possessor of genius; genius is an inheritance. Kivy believes that even though we can get tongue- or thought-tied when we try to define the term, genius refers; that is, the term genius is about something and it is possible to identify geniuses. He is, however, driven by the myth of individualism and so regards any effort to overturn the theory of genius in terms of "possessed" or "possessor" as by definition reductionist. For this reason he dismisses with high praise Battersby's feminist critique which clearly grounds the very idea of genius in patriarchal and patrilineal traditions.

Thinking in terms of what Kivy describes with prejudice as "socio-political deconstruction," and guided by feminist perspectives, Battersby offers what is essentially a sociological analysis of the origins and nature of genius. Her analysis can be brought to bear on contemporary efforts to bring more women and minorities into the fields of science and technology. The STEM paradigm reinforces masculine and patriarchal views about science and technology as the only fixes we need to solve problems whether they are technological or socio-cultural. Indeed, from the perspective of the "technological fix," all problems reduce to technological problems or problems that can be solved by the physical and natural sciences. The dangers here can be brought into focus if we consider why (with some recent exceptions) the arts, social sciences, and humanities have been left out of STEM.

Sociology is anathema to philosophers like Kivy who see sociological analyses (along with analytic traditions in the arts and humanities) as by definition reductionist and not scientific. Kivy is dedicated to protecting the individuality of the genius and claims contrary to all the evidence that there is no natural barrier to women becoming geniuses. Of course the barrier is social and cultural and therefore invisible to Kivy who clearly suffers from dissocism.

Once we understand with Battersby that "I am a man," "I am a genius," "I am a God" are one statement we can understand the trouble Gertrude Stein caused when she proclaimed "I am a genius." On the complexity of Stein's relationship to the ideas of genius, the world at large, the world of family and household, and the world of the writer see (Perelman 1994: 129–169). Whatever the problems with operationalizing the concept of genius, there can be no question that it has historically and culturally been an appellation unavailable to and inapplicable to women. Women have not even benefitted from the illusion of the "I" which has a masculine gender.

We can get a clearer view of Kivy's dissocism by reviewing his critique of sociologist Tia DeNora's *Beethoven and the Construction of Genius* (1995). This debate reveals with greater clarity than the debate between Kivy and Battersby the traditional warfare between sociology and philosophy. In the case of Kivy and Battersby we have two philosophers at odds. Battersby has taken the sociological turn; Kivy represents the philosopher as a defender of "common sense" views of the individual. In his debate with DeNora, common sense individualism is opposed to a view of the self as social and embedded in networks.

We can see the problem in a nutshell by noting Kivy's confusion about the scare quotes DeNora puts around "own" in the phrase "his 'own' talent." Kivy's dissocism blocks him from recognizing that what is operating in that phrase is a sociological concept of the self as a social fact, a social structure. We begin to see in Kivy's case a demonstration of the role that privilege plays in generating dissocism.

Common sense is nothing more than the prevailing accumulated institutionally grounded understanding of the everyday taken for granted world. It is not a sophisticated philosophical, scientific, or logical perspective that can ground arguments. It is common sense to view Beethoven as possessing genius, as a born natural genius; he is not the kind of genius who is possessed by spirits, gods, or angels. Both kinds of genius are grounded in the myth of individualism. Kivy cannot read DeNora because he is blinded by dissocism; he cannot see the social. I have been at pains throughout this book to stress that sociological analysis does not deny the uniqueness of Einstein, Beethoven, Mozart, or any of those whose achievements have earned them the label "genius." The sociological claim is that uniqueness is not a simple function of genes, neurons, or more mysterious "gifts," but rather a function of the uniqueness of the social networks through which a person's life unfolds.

Genius: The Very Idea

The very idea of "genius" is loaded with dangers and it might be wise to eliminate it from our vocabulary. The most immediate danger of course is that it fuels the myth of individualism. If we consider its etymology we find it associated with a spirit present at birth, innate ability, and divine inspiration. The latter idea comes from the original meaning of the term: a tutelary god, moral guide, or spirit attendant on a person. Classically, the genius is guided from birth on by a guardian deity. By the mid-seventeenth century it had come to mean exceptional natural ability.

The term has been the subject of much debate (is it at all meaningful?) And the debate has tended toward resolution by reducing the term to more readily operationalizable ideas like creativity and eminent achievement. It is no accident that the term "genius" achieves increasing prominence from the 1500s onward coincidentally with the emergence of the capitalist model of an economy based on private property and individual (to the point of atomistic) behavior in the market. Its origins go back, of course, to ancient Rome. Mentions of the term appear to have been in decline since the 1800s.

Genius allows us, if we wish, to erase the genius cluster and the multiples so that an invention or discovery can be assigned to a particular individual. To recall the implication of the concept of the social self, individuality is not a matter of what is inherent, genetic, or neuronal but rather of the particular social configuration that represents the social groups and networks that mark the self's progress through the life span. To put it in the most radical terms, following the polish sociologist Gumplowicz, we can say that the individual is a voice box for those groups and networks; from the perspective of the sociology of knowledge we can say that ideas are generated by and in social networks and manifest themselves in the cognitive experiences of individuals who can experience thoughts and collectively construct technologies that manifest those thoughts.

Finally, looking to Galton's (1822–1911) work on hereditary genius, we should not ignore the connection of the genius literature from Galton on with the eugenics movement. The myth of genius lies in its connection to the myth of individualism. I have noted that the myth of individualism is fueled by the ethos of capitalism. The relationship between the coincident emergence of the capitalist ethos and the modern concept of genius has been noted by others. So has the relationship of the modern concept of genius to the emergence of new ideas about aesthetics and the self.

Individual, Genius, and Social Context

My approach until now has been to approach the critique of the very idea of "genius" through the lens of sociology. Suppose we approach the critique through the lens of psychology. What can we say about the viability of the very idea of an individual genius? There is a social process that transforms originally non-intuitive ideas into intuitively viable ideas. Non-intuitive ideas are ideas that engage new experiences. New challenging experiences (the heliocentric solar system, relativity theory, the germ theory of disease) if they bear the burden of evidence over time become absorbed into everyday institutions and become part of our reservoir of ideas that are "common sense" and intuitively accessible (Restivo 1992: 121). Intuitions are as yet unpacked processes that accumulate over time and come together in insights (flashes of insight) that appear to come out of nowhere. To have an intuition is to have a thought whose origins you are not aware of because you do not have access to the unfolding of your thoughts (the fallacy of introspective transparency; and see Weisberg 1993: 56–57).

Simonton (1999: 26–27) drawing on Campbell's (1960) ideas on blind variation and selective retention, proposed that new ideas emerge as the result of "chance permutations." This view proposes that "original ideas" are not grounded on the individual thinker's previous ideas and associations. This idea is contradicted by laboratory studies of problem solving (never mind for the moment the sociological perspective on social networks as the loci of ideas).

The psychology of genius is based on the assumptions that (1) "genius" can be measured, say the way we measure IQ; that is, that it is a trait; (2) the unique characteristics of the personalities of creative individuals are causally related to creativity; for example, this might lead us to believe that if creative individuals are more "autonomous" than non-creative individuals, autonomy might be a causal factor in creativity; the same can be said for madness. The assumption is subject to the fallacy of "post hoc ergo propter hoc"; and (3) "possessing genius" is an enduring trait. The critical literature on IQ as a trait is enormous. It comes down to recognizing that one number cannot summarize the intellectual differences between people (Hampshire et al. 2012).

The evidentiary grounds for the assumption that madness and genius are connected are weak (Weisberg 1993: 70ff.). The association between creativity and mental illness tends to be rooted in remarks by historical sages. We moderns still feel obliged to draw on ancient philosophers, notably Socrates, Plato, and Aristotle to guide and support our research. There is, in any case, a continuing debate about genius and madness and the implications of this "connection" for creative persons and psychiatric patients. The issue is discussed in Plato's dialog, Phaedrus. The problem, as Socrates sees it, is to characterize madness scientifically and not rhetorically. Socrates identifies two categories of mental illness: those that arise from a biological disease; and those that are deviations in behavior that violate prevailing norms of conduct. These are the non-conformists Plato described as "divinely inspired."

The link between genius and divine madness has been discussed for thousands of years; its association with clinical madness is a modern invention (Becker 2000, 2014). The research on creativity does not support the idea that "geniuses" or creative thinkers possess unique cognitive abilities or thought processes. The significant differences between creative and non-creative individuals include levels of persistence, motivation, and dedicated work protocols; in addition, there are the social and cultural forces and contexts that I stress as a social scientist.

"Genius" is a social category. It is in this sense a relative attribution and not an absolute trait. If genius were a trait, the evaluation of geniuses would not change substantially over time. And geniuses would be consistently creative. There are, however, many well-known counter-examples. Bach's "genius" went unrecognized (or better he was not so labeled) until relatively recently; his music was ignored for decades following his death. The Dutch painter Jan Lievens was more esteemed in his lifetime than Rembrandt. How many of Lievens' paintings have you seen?

The novelty of a creative achievement, its utility, and the genius that accrues to the creator is dependent on context. Whitney's cotton gin was considered a work of genius because it became embedded in a cotton economy and led to newer developments. The invention would not have led to acclaim in a society without a cotton economy or with a minor one. This is somewhat ingenuous and indeed moot if indeed necessity is the mother of invention. How likely is it that Whitney would have worked on developing a cotton gin in a non-cotton economy? The salient point is that social context was *a* if not *the* key to labeling Whitney a genius. Furthermore, Whitney's invention had artifactual antecedents (Parayil 1999: 31–32; and see Basalla 1988). Mechanical cotton gins were already in use in the South before Whitney's invention, and early versions were used in India more than 2000 years ago, as early as the twelfth century in Italy and the fourteenth century in China.

There are numerous examples, then, of individuals who are ignored then lauded or lauded and then ignored. Einstein's "genius" was not demonstrated when it came to the development of quantum mechanics in the same way it was with relativity theory. One might want to argue that Einstein will eventually prove to have been right and quantum mechanics will develop a deterministic framework. But we can turn to other areas that Einstein paid attention to during his lifetime and in which he did not exhibit "genius," including the history of science, politics, and religion. There are no "universal geniuses."

There are in the wider literature on genius allusions to intuition. Terence Tao, the mathematician who won the Fields Medal in 2006, argues that what really matters in creative achievement is "hard work, directed by intuition, literature, and a bit of luck." Watson and Crick are reported to have relied on intuitive non-logical leaps in their quest for the structure of DNA (Adams 1979: 61–62). And indeed, intuition is one of the personality characteristics reported for geniuses (Weisberg 1986: 73; and see Simonton 1999: 32–33, 83ff.; and on the contrast between intuitions and ordinary perceptions see Mercier and Sperber 2017: 63–67).

Uncovering the Mystery of Intuition

Intuition is generally understood to mean unmediated, direct perception or apprehension. It is, in myth, pure, unlearned, and not logical. Moreover, it is understood to more often than not access the truth of a situation. The Meyers-Briggs MBTI personality type index describes a person who "reads between the lines" to determine meaning, thinks in "leaps," explicitly shows an interest in thinking and behaving in novel ways, sees "the big picture" before pursing the facts, is more motivated by impressions, symbols, and metaphors than experiences, and is more oriented to thinking about novel possibilities than in how to realize them in reality. Meyers-Briggs categorizes such as a person as "the Architect," Introverted-Intuitive-Thinking-Perceptive (INTP). Intuitions are automatic (but as we will soon see, all thinking may be automatic), tacit, immediate, affective, and directed to something specific.

Prior to the emergence of theories of automaticity and autopoiesis (see below) students of cognitive processes recognized the existence of certain hard-wired processes. The tendency to attend to faces and to see faces in incomplete or abstract objects is arguably the primary example of the brain's capacity for "intuition." Intuition is also grounded in our experience of our own consciousness which leads us to believe that if we aren't being attentive we aren't thinking. This turns out to be very far from the truth.

The brain is highly interconnected and is governed by a dynamical systems (chaotic) energy that is always "on," working in the background drawing connections between the cultural capital we have accumulated in its store of information. This chaotic energy can be thought of as the basal neurological metabolism of the brain. It is the motor that keeps the brain's connections active at different levels. These levels represent different levels of consciousness and awareness. Under normal circumstances, unconscious work is elevated to our attention situationally and when this happens "suddenly" it is experienced as an "intuition." Sometimes the process is slower; for example, we may go to sleep after giving up on solving a math problem and after waking up slowly come to realize that we have the solution. We think of intuition as non- or irrational and visceral but it is one of the rational ways the brain works. I need to stress when I write about "the brain" in this way I am not writing about the classical autonomous brain that is not connected to the body and the world.

Is there an evolutionary advantage to intuition? Perhaps. It allows us to act/think without a complete understanding of the motivating reasons. It can serve us well in situations that require fast and immediate action; sometimes, this is just a cognitive moment that gives insight but does not necessarily provoke or require action. This is referred to in some decision making circles as Recognition Primed Decision strategy (Klein 1998: 1–30). The drawbacks of intuitions are that they are almost by definition incomplete, error prone, and virtually impossible to communicate to others except in the most superficial terms. They are subject to unconscious biases that can creep in when we in a sense think and act before we have thought. There is a more radical and counter-intuitive version of this idea and that is we never think in a willful sense; we "become aware" of thoughts generated by the brain/mind/culture/world system. To put this in the form of a slogan, "thoughts are after thoughts." Thoughts are generated at the intersection of the stream of affordances and the stream of consciousness. Affordances are the sea of sensory inputs we engage as we move through our environments (see notes). There are a number of provocations for the claim that thoughts are afterthoughts.

Classically, and even in our own era, it has been widely assumed in lay and scientific circles that we humans consciously, systematically, and rationally think and act on the basis of our "willful" thinking. Langer (1978) and Langer, Blank, and Chanowitz (1978) questioned this assumption and more recently we have witnessed the emergence of a research literature on "automaticity" (Bargh and Chartrand 1999: 462). The three major forms of automaticity are "an automatic effect of perception of action," "automatic goal pursuit," and "a continual automatic evaluation of one's experience." From the accumulating evidence, the authors conclude that these various nonconscious mental systems perform the lion's share of the self-regulatory burden, beneficently keeping the individual grounded in his or her current environment. There is a history of two-stage models of free will that have been discussed by writers from William James to Karl Popper and Daniel Dennett to Roger Penrose (http://www.informationphilosopher.com/freedom/two-stage models.html); and see Campbell's "blind variation & selective retention" two-stage model for free will: first chance, then choice; first "free," then "will." Thoughts *come to us* freely. Actions *go from us* willfully.

These two-stage models are designed to save free will but reflect the fallacy of introspective transparency, ignorance of automaticity, and a failure to account for or to even acknowledge social and cultural programming.

Conclusion

Let's return once more to question I opened this book with: why another book on Einstein? Recall the Kalb article I discussed in Chap. 1 in which Einstein is described as creating relativity theory ab novo out of his own individual mind. Awe inspiring "genius" rooted in genes and neurons has been the default "explanation" for Einstein for decades. Einstein, like all of us, was the beneficiary of the human capacity to experience events at the intersection of the stream of affordances and the stream of consciousness. Both streams are fueled by the social and ecological contexts of our lives as they unfold by way of interaction rituals and interaction ritual chains (see Chap. 6) in the social networks we traverse. To put it dramatically, we are all like Tolstoy's "genius" general Napoleon, blind instruments of history.

Appendix: A Note on Creativity and Madness

> Men have called me mad; but the question is not yet settled, whether madness is or is not the loftiest intelligence-whether much that is glorious-whether all that is profound-does not spring from disease of thought-from moods of mind exalted at the expense of the general intellect: They who dream by day are cognizant of many things which escape those who dream only by night. In their grey vision they obtain glimpses of eternity. ... They penetrate, however rudderless or compassless, into the vast ocean of the "light affable". Edgar Allan Poe (1809–1849) "Eleonora," 1903/1842.

The lunatic, the lover, and the poet are of imagination all compact: Theseus, A Midsummer Night's Dream (Act V, Scene 1, Shakespeare 1998/1600).

> Marcel Proust, "Everything great in the world is created by neurotics. They have composed our masterpieces, but we don't consider what they have cost their creators in sleepless nights, and worst of all, fear of death: The Guermantes Way," pt. 1, *Remembrance of Things Past*, vol. 5 (1921).

The idea that there was a connection between genius and madness was already afoot in the ancient world. Aristotle, in Problemata XXX.1, 953a, 10–14 (Barnes 1984), posed the following question: "Why is it that all those who have become eminent in philosophy or poetry or the arts are clearly melancholics and some of them to such an extent as to be affected by diseases caused by black bile?" The influence of Cesare Lombroso's

(1891) belief that genius and madness were connected was still strong when Lewis Terman's (1925) data suggested that people of high ability exhibited less incidence of mental illness and adjustment problems than average (Neihart 1998; and for a collection of contemporary longitudinal studies of giftedness and talent, see Subotnik and Arnold 1994). On the genius-madness connection, see Wolchover (2012), Marano (2007), Schlesinger (2009), and Kyaga (2015). Kay Redfield Jamison (1995) found that outstanding 16-year-olds were more likely to become bipolar in later life than their more "normal" peers (and see her related work: 1989, 1993). Fallon has found that the same neural circuits are associated with bipolar disorders and creativity, reported in Hsu (2012); and see Fallon (2013). Saks (2007) explains that psychotics can hold contradictory ideas in mind and access loose associations that most people's "brains" wouldn't allow to emerge into the conscious mind.

Some writers on the creativity-madness connection stress the hard work and late nights characteristic of creative people; see the remarks by M. Csikszentmihalyi and Robert Root-Bernstein in Marano (2007); and see Csikaentmihalyi (1996) on flow and creativity, and Macarthur award recipient Root-Bernstein's (1989) virtuoso examination of the conditions most likely to lead to discoveries in science; his ideas converge on the anarchistic theory of inquiry I outlined in Restivo (2016).

There is a sociology of genius/creativity and madness that undercuts genetic and neuronal explanations. The creative lifestyle is not conducive to emotional stability and many creative people have to deal with poverty and public indifference. And alternatively there are social pressures that accompany success and public acclaim. Creative people are likely to be more open and sensitive and thus more easily exposed to suffering and pain. We are at the end of the day left to deal with contradictions. Koh (2006), in her review of the link between creativity and madness, points out that there is indeed substantial evidence linking genius and madness but that similar factors may be at work in "geniuses" and "ordinary" people. Furthermore, post-modern culture may have erased the fine line between genius and madness. As recently as 1993 Natalie Angier reported in the New York Times that in the wake of decades of controversy about how to define "madness" and "creativity," and resistance by scientists to the popular idea that the two terms reflect a relationship, there is now powerful evidence linking certain mental disorders and artistic achievement. Critics of these findings include Weisberg (1986, 1993); and Rothenberg's (2014) conclusions are based on interviews with 45 Nobel laureates, hardly a representative sample capable of carrying the burden of a firm conclusion on the link between creativity and madness; but see also his (1990). Even though Nobelist Eric R. Kandel (2018) is still stuck in the pseudo-problem of how mind emerges from physical processes, his study of the biology of "unusual minds" and what they reveal about ourselves is based on a lifetime of outstanding scientific achievement and worth the attention of anyone interested in the brain, brain disorders, and creativity; see especially Chap. 6 on "Our Innate Creativity: Brain Disorders and Art." He discusses "psychotic art" (the visual art of people with schizophrenia), and the creativity of people with bipolar disorder, autism, Alzheimer's disease, and frontotemporal dementia. Chapter 2 is on "Our Intensely Social Nature: The Autism Spectrum." Kandel does not explore the implications of the social brain beyond the biological paradigm thus failing to follow up on "our intensely social nature."

Bibliographic Notes for Chapter 4

Genius Clusters and Multiple (Simultaneous) Discovery

It's clear from Claudia Kalb's (2018) work in "The Science of Genius" that the concepts of social networks of geniuses and genius clusters are known to students of genius. The problem is that networks and clusters are viewed as contexts or environments that are associated with the genius (see Chap. 3).

On genius and culture and genius clusters, see the anthropologist Kroeber (1963, esp. 7–27 and 838–846); for a more journalistic and accessible account, see Weiner (2016); major research is reported in Simonton (1999: 199–241), Mercier and Sperber (2017: 315–327); the locus classicus for the concept of multiple discovery is Merton (1961).

The Magic and Mystery of Mathematics

For a journalistic review, see Browne (1987). John Dee was arrested for "calculating" in 1555: see French (2002/1972); Thorndike (1923–1958); Zetterberg (1980). St. Augustine (354–430), in *The Confessions* (1960/397–400: 95; 116; 163; 241), warned Christians against "mathematicians" (read "astrologers," "prophets," "necromancers," "numerologists"); all those who traded in numbers had made a pact with the devil. See also Literal Commentary on Genesis (*De Genesi and litteram*, an exegetical text on the first book of the Bible dating between 1147 and 1164 (see Book II, xviii, 37).

Non-Euclidean Geometry, Ramanujan, Nikola Tesla, Rodin

See Restivo (2018: 139–142) and the associated references; on "having a feeling for the organism," see Keller (1983); the movie clip on Rodin is from the film *Camille Claudel*, directed by Bruno Nuytten for Les Films Christian Fechner, 1988.

1840–1920 Genius Cluster

See Restivo (2018), on the central period of the social science Copernican revolution, 1840–1920/1930; on the historical anthropology and configurations of genius clusters, see Kroeber, 1963); Weiner (2016) is an excellent journalistic history and companion to Kroeber; and for the late nineteenth and early-twentieth-century genius cluster that nourished Einstein, see Miller (2001); and also see Kalb (2018: 77–112).

Gender and Genius (+Queer Genius)

Battersby (1989); Elfenbein (1996); Stadler (1999); Hone (2015); DeNora (1993); on Kivy vs. Battersby & Kivy vs. DeNora, see Kivy (2001); Garber (2018); for a feminist perspective on mind, brain, and cognition, see Wilson (1998).

Genius: The Very Idea

On the history of genius, see McMahon (2013); and Martin (1969); on the scientific revolution which helped fuel the evolution of the concept of "genius", see Restivo, 1994: 29–48); on "Darwinian Genius: The Future of an Idea," see Simonton (1999: 243–248); for a study of genius in the general context of reason, see Mercier and Sperber (2017); Jenkins (2013) urges us to place genius in an open, democratic space; this, he writes will force us to revolutionize the way we think of "knowledge," "school," and "texts." This will help us re-imagine various forms of life as "spaces of significant learning." In this context, different forms of work from songs and poetry to performances and videos can be viewed as critical texts. Hip-hop is a "space of knowledge production" that expands the boundaries of academic inquiry and transforms our understanding of and approach to pedagogy. Marjorie Garber (2000) deals with "genius" as a contemporary commodity, ambition, and lifestyle. She writes that biographers, scholars, critics, and fans are trying to nail down a concept that can't be nailed down and simultaneously trying to "make the genius lovable, accessible, and ready for prime time." There is a journalistic truth here but one that skirts around the program for a sociology and anthropology of genius that I have argued for in this book. She reviews the word "genius," "genius and aberration," "quantifying genius" (on Terman's efforts to identify genius using a version of Binet's IQ test), "banking on genius" (attempting to foster genius through grants such as the "genius grants" of the MacArthur Foundation and the Repository for Germinal Choice, the "genius" sperm bank founded by R.K. Clark), and "our genius complex"; unless we separate the power of ideas from personalities we will continue to be "dazzled" by celebrity and distracted by the lauding of geniuses as high-culture heroes, as "essence rather than force." I endorse her call for a new way of thinking about thinking rather than just another word. For readable biographies of "geniuses," see Gleick (1992) on Feynman; Monk (1990) on Wittgenstein, an outstanding biography that can be paired with Janik and Toulmin (1973) which is a virtually sociological account that situates Wittgenstein in the Vienna of his time; see also Jamison's (2017) study of "genius, mania, and character" in the life of Robert Lowell; and on "idiot savants," see Howe (1989), and for his explanation of "genius," see Howe (1999).

Individual Genius and Social Context, Uncovering the Mystery of Intuition, and Free Will Redux

On "automaticity," see Bargh and Chartrand (1999); Langer (1978); and Klein (1998); Weisberg (1993) is an excellent treatment of the genius myth; he was already exposing the myth of genius in his 1983 book; and see Maturana and Varela (1980): They claim that living systems are "autonomous, self-referring and self-constructing closed systems—in short, autopoietic systems" (R.S. Cohen and M.W. Wartofsky, Editorial Preface, v). The approach here is based on a virtually essentialist biological cognition paradigm. It does afford, however, opportunities to link it to the social brain connectome paradigm I discuss in Chap. 5.

Norretranders (1991) has an innovative and somewhat off-beat conception of consciousness. He claims that we have oversold consciousness: it does not contain much information, "for information is otherness and unpredictability." "The more power consciousness has over existence, the greater the problem of its paucity of information becomes." "Consciousness will find composure by acknowledging that people need more information than consciousness can supply. Man also needs the information contained in consciousness, just as we need a map to find our way around the terrain. But what really counts is not knowing the map—it is knowing the terrain" (416–417). This gives only the slightest hint at the eccentric riches contained in this big bold book. I should note his indebtedness to the works of Benjamin Libet. On intuition, see Lynch (2006), Bealer (1998); Klein (2003); Giannini, Daood, et al. (1978); Robson and Miller (2006); DePaul and Ramsey (1998); Kahneman (2011). There is a large literature on intuition in eastern philosophy; see, for example, Radhakrishnan (1932) for a review of the concept in Eastern and Western philosophy and a defense of intuition, faith, spiritual experience, and the testimony of scriptures in theological language as "necessary for knowledge and life." For contemporary perspectives in this tradition, see Chopra and Orlogg (2001) and Aurobindo (1990).

On Affordances

I have borrowed and adapted the term "affordances" from an admittedly already complicated literature on the term. Gibson (1986) introduced the term and defined it as "action possibilities in the environment." Norman (1988) introduced the term in the Human Computer Interaction literature and defined affordances as "perceived properties that may or not actually exist." Gaver separates affordances from their perceptive qualities (see the reviews in Soegaard (2015) and Kaptelinin (2014)). Gaver (1991) tried to clarify the various ambiguities of the term; also see Gaver (1996).

Why is the conversion factor in the energy-mass equation c^2?

Rathore (2020).

References

Adams, J.L. 1979. *Conceptual Blockbusting*. 2nd ed. New York: Norton.
Aurobindo, S. 1990. *Synthesis of Yoga*. Twin Lakes, WI: Lotus Light Publications.
Bargh, J.A., and T.L. Chartrand. 1999. The Unbearable Automaticity of Being. *American Psychologist* 54 (7): 462–479.
Barnes, J., ed. 1984. *The Complete Works of Aristotle*. Vol. II. Princeton: Princeton University Press.

Bartocci, U. 1999. *Albert Einstein & Olinto De Pretto*. Bologna: Societa Editrice, Andromeda.

Basalla, G. 1988. *The Evolution of Technology*. Cambridge: Cambridge University Press.

Battersby, C. 1989. *Gender and Genius: Towards a Feminist Aesthetics*. Bloomington: Indiana University Press.

Bealer, G. 1998. Intuition and the Autonomy of Philosophy. In *Rethinking Intuition: The Psychology of Intuition and Its Role in Philosophical Inquiry*, ed. M. Depaul and W. Ramsey, 201–239. Lanham, MD: Rowman & Littlefield.

Becker, G. 2000. The Association of Creativity and Psychopathology: Its Cultural-Historical Origins. *Creativity Research Journal* 13 (1): 45–53.

———. 2014. A Socio-Historical Overview of the Creativity-Pathology Connection from Antiquity to Contemporary Times. In *Creativity and Mental Illness*, ed. J.C. Kaufman, 3–24. Cambridge: Cambridge University Press.

Browne, M.W. 1987. Mathematics and Magic. *The New York Times Magazine*, October 18 (nytimes.com).

Campbell, D.T. 1960. Blind Variation and Selective Retention in Creative Thought as in Other Knowledge Processes. *Psychological Review* 67: 380–400.

Chopra, D., and J. Orloff. 2001. *The Power of Intuition* (Audio CD). New York: Hay House.

Csikaentmihaly, M. 1996. *Creativity: Flow and the Psychology of Discovery and Invention*. New York: Harper Perennial.

DeNora, T. 1993. *Beethoven and the Construction of Genius: Musical Politics in Vienna, 1792–1803*. Berkeley: University of California Press.

DePaul, M., and W. Ramsey. 1998. *Rethinking Intuition: The. Psychology of Intuition and its Role in Philosophical Inquiry*. Lanham, MD: Rowman & Littlefield.

Elfenbein, A. 1996. Lesbianism and Romantic Genius: The Poetry of Anne Bannerman. *English Literary History* 63 (4): 929–957.

Fallon, J. 2013. *The Psychopath Inside: A Neuroscientist's Personal Journey into the Dark Side of the Brain*. New York: Current/Penguin.

French, P.J. 2002. *John Dee: The World of the Elizabethan Magus*. New York: Routledge; reprint of the 1972 Ark Paperback.

Garber, Megan 2018. David Foster Wallace and the Dangerous Romance of Male Genius on the "Centrifugal Forces of Talented Men". *The Atlantic*, May 9. https://www.theatlantic.com/entertainment/archive/2018/05/the-world-still-spins-around-male-genius/559925/.

Garber, Marjorie. 2000. Our Genius Problem. *The Atlantic*, December, 2002: 65–72. https://www.theatlantic.com/magazine/archive/2002/12/our-genius-problem/308435/.

Gaver, W.W. 1991. Technology Affordances. In *Proceedings of the CHI*, 79–84. New York: ACM Press.

———. 1996. Affordances for Interaction: The Social Is Material for Design. *Ecological Psychology* 8 (2): 111–129.
Giannini, A.J., J. Daood, et al. 1978. Intellect versus Intuition-Dichotomy in the Reception of Nonverbal Communication. *Journal of General Psychology* 99: 19–24.
Gleick, J. 1992. *Genius: The Life and Science of Richard Feynman*. New York: Vintage.
Hampshire, A., R.R. Highfield, et al. 2012. Fractionating Human Intelligence. *Neuron* 76 (6): 1225–1237.
Hone, M. 2015. Gay Genius: From Plato to Nietzsche to Byron (© Michael Hone).
Howe, M.J.A. 1989. *Fragments of Genius: The Strange Feats of Idiot Savants*. London: Routledge.
———. 1999. *Genius Explained*. Cambridge: Cambridge University Press.
Hsu, C. 2012. Scientists Find Truth in "Mad Scientist" Stereotype: There Is a Link between Genius and Insanity. Medicaldaily.com, June 4.
Jamison, K.R. 1995. *An Unquiet Mind*. New York: Alfred A. Knopf.
———. 2017. *Robert Lowell: Setting the River on Fire*. New York: Alfred A. Knopf.
Janik, A., and S.E. Toulmin. 1973. *Wittgenstein's Vienna*. New York: Simon & Schuster.
Jenkins, T. 2013. De (Re) Constructing Ideas of Genius: Hip Hop, Knowledge, and Intelligence. *International Journal of Critical Pedagogy* 4 (3): 11–23.
Kalb, C. 2018. *The Science of Genius*. National Geographic: Special Publication.
Kahneman, D. 2011. *Thinking, Fast and Slow*. New York: Farrar, Strauss, and Giroux.
Kandel, E.R. 2018. *The Disordered Mind: What Unusual Brains Tell Us About Ourselves*. New York: Farrar, Strauss, and Giroux.
Kaptelinin, V. 2014. *Affordances and Design*. Aarhus N, Denmark: The Interaction Design Foundation.
Keller, E.F. 1983. *A Feeling for the Organism: The Life and Work of Barbara McClintock*. New York: W.H. Freeman.
Kyaga, S. 2015. *Creativity and Mental Illness: The Mad Genius in Question*. London: Palgrave Macmillan.
Kivy, P. 2001. *The Possessor and the Possessed: Handel, Mozart, Beethoven and the Idea of Musical Genius*. New Haven: Yale University Press.
Klein, G.A. 1998. *Sources of Power: How People Make Decisions*. Cambridge, MA: MIT Press.
———. 2003. *Intuition at Work*. New York: Random House.
Koh, C. 2006. Reviewing the Link Between Creativity and Madness: A Postmodern Perspective. *Educational Research and Reviews* 1 (7): 213–221.
Kroeber, A. 1963. *Configurations of Culture Growth*. Berkeley: University of California Press.

Langer, E., A. Blank, and B. Chanowitz. 1978. The Mindlessness of Ostensibly Thoughtful Action: The Role of "Placebic" Information in Interpersonal Interaction. *Journal of Personality and Social Psychology* 36: 635–642.

Lombroso, C. 1891. *The Man of Genius*. London: Walter Scott.

Lynch, M. 2006. Trusting Intuitions. In *Truth and Realism*, ed. P. Greenough and M. Lynch, 227–238. Oxford: Oxford University Press.

Martin, P., ed. 1969. *Genius: The History of an Idea*. New York: Basil Blackwell.

Maturana, H.R., and F. Varela. 1980. *Autopoiesis and Cognition*. Dordrecht: D. Reidel.

Marano, H.F. 2007. Creativity and Mood: The Myth that Madness Heightens Creative Genius. https://www.psychologytoday.com/us/articles/200705/genius-and-madness?quicktabs_5=1.

McMahon, D.M. 2013. *Divine Fury: A History of Genius*. New York: Basic Books.

Mercier, H., and D. Sperber. 2017. *The Enigma of Reason*. Cambridge, MA: Harvard University Press.

Merton, R.K. 1961. Singletons and Multiples in Scientific Discovery: A Chapter in the Sociology of Science. *Proceedings of the American Philosophical Society*, 105: 470–486; reprinted in R.K. Merton (1973). *The Sociology of Science*. Chicago: University of Chicago Press, 1973, pp. 343–370.

Miller, A.I. 2001. *Einstein, Picasso: Space, Time, and the Beauty That Causes Havoc*. New York: Basic Books.

Monk, R. 1990. *Ludwig Wittgenstein: The Duty of Genius*. New York: Penguin.

Neihart, M. 1998. Creativity, the Arts, and Madness. *Roeper Review* 21: 47–50.

Nietzsche, F. 1996/1878–1880. *Human, All Too Human*. Cambridge: Cambridge University Press.

Norman, D. 1988. *The Psychology of Everyday Things*. New York: Basic Books.

Norretranders, T. 1991. *The User Illusion: Cutting Consciousness Down to Size*. New York: Viking.

Parayil, G. 1999. *Conceptualizing Technological Change*. Lanham, MD: Rowman & Littlefield.

Perelman, B. 1994. *The Trouble With Genius: Reading Pound, Joyce, Stein, and Zukovsky*. Berkeley: University of California Press.

Radhakrishnan, S. 1932. *Idealist View of Life*. New York: Simon & Schuster.

Rathore, H. 2018. Why Did Einstein Use Speed of Light Squared in the Famous Equation $E = mc^2$? https://www.quora.com/Why-did-Einstein-use-speed-of-light-squared-in-the-famous-equation-E-mc-2.

Restivo, S. 1992. *Mathematics in Society and History*. New York: Springer.

———. 1994. *Science, Society, and Values: Toward a Sociology of Objectivity*. Bethlehem, PA: Lehigh University Press.

———. 2016. *Red, Black and Objective*. New York: Routledge.

———. 2018. *The Age of the Social: The Discovery of Society and the Ascendance of a New Episteme*. New York: Routledge.

Robson, M., and P. Miller. 2006. Australian Elite Leaders and Intuition. *Australasian Journal of Business and Social Inquiry* 4 (3): 43–61.

Root-Berstein, R. 1989. *Discovering, Inventing and Solving Problems at the Frontiers of Science*. Cambridge, MA: Harvard University Press.

Rothenberg, A. 1990. *Creativity & Madness*. Baltimore, MD: Johns Hopkins University Press.

———. 2014. *Flight from Wonder: An Investigation of Scientific Creativity*. New York: Oxford University. Press.

Saks, E. 2007. *The Center Cannot Hold: My Journey Through Madness*. New York: Hyperion.

Schlesinger, J. 2009. Creative Myth Conceptions: A Closer Look at the Evidence for the 'Mad Genius' Hypothesis. *Psychology of Aesthetics, Creativity, and the Arts* 3 (2): 62–72.

Shakespeare, W. 1998/1600. A Midsummer Night's Dream (Act V, Scene 1, openinglines).https://www.gutenberg.org/files/1514/1514-h/1514-h.htm.

Simonton, D.K. 1999. *Origins of Genius*. New York: Oxford University Press.

Soegaard, M. 2015. Affordances. In *The Glossary of Human Computer Interaction*, ed. B. Papantoniou, M. Soegaard, J. Lupton, et al. Aarhus N, Denmark: The Interaction Design Foundation.

Stadler, G. 1999. Louisa May Alcott's Queer Geniuses. *American Literature* 71 (4): 657–677.

Subotnik, R.F., and K.D. Arnold. 1994. *Beyond Terman*. Norwood, NJ: Ablex Publishers.

Terman, L. 1925. *Mental and Physical Traits of a Thousand Gifted Children*, Vol. 1 of 5 in Genetic Studies of Genius. Stanford: Stanford University Press.

Thorndike, L. 1923–1958. *A History of Magic and Experimental Science*, 8 vols. New York: Macmillan & Columbia University Press.

Weiner, E. 2016. *The Geography of Genius: Lessons from the World's Most Creative Places*. New York: Simon & Schuster.

Weisberg, R. 1986. *Creativity: Genius and Other Myths*. New York: W.H. Freeman.

———. 1993. *Creativity: Beyond the Myth of Genius*. New York: W.H. Freeman.

Wilson, E.A. 1998. *Neural Geographies*. New York: Routledge.

Wolchover, N. 2012. Why Are Genius and Madness Connected. https://www.livescience.com/20713-genius-madness-connected.html.

Zetterberg, J.P. 1980. The Mistaking of 'the Mathematicks' for Magic and Tudor and Stuart England. *Sixteenth Century Journal* 11: 83–97.

CHAPTER 5

The Social Brain Paradigm

Abstract This chapter introduces my perspective on and my model of the social brain. The development of the social brain paradigm reflects a general development from hierarchical to network thinking across the intellectual spectrum during the latter part of the twentieth century. I discuss the evolution of the social intelligence hypothesis into the social brain hypothesis, and the reigning myths about the brain that have obstructed social brain thinking. I review the key developments in the history of neuroscience at its nexus with the life- and social sciences and their connections to social brain research and theory. The chapter ends with a presentation of my model of the social brain as a networked information system situated in and coupled with a social ecology (Appendix 1). In Appendix 2, I review the concept of connectomics, and in Appendix 3, I list links to glossaries on brain terminology to aid readers in understanding the terms used in the text to describe the structure and function of the brain.

Keywords Social intelligence • Social brain • Connectomics • Information

S. Restivo, *Einstein's Brain*,
https://doi.org/10.1007/978-3-030-32918-1_5

From Hierarchies to Networks

Against the background of the previous chapters, I can now outline a new concept of the brain that eliminates "brains in a vat" thinking and the neuroist idea that the brain is the originating locus of all of our thoughts and behaviors. This "new" concept is not so much new as it is an effort to integrate the various emerging proposals about the need to revise classical thinking about the brain/mind/body and the directions such proposals should be taking.

Hierarchical models have dominated approaches to the modern understanding of cognition at least since the contributions of Jean Piaget. Indeed, hierarchical thinking guided model and theory building across the entire spectrum of the sciences throughout most of the twentieth century. During the second half of the twentieth century, postmodernist thinkers led or provoked the struggle against classical dichotomies, positivist science, and philosophies grounded in the concept of foundations. This struggle undermined categories and classifications that had reigned for millennia in some cases and for hundreds of years in others. Male and female, life and death, nature and nurture are just among the most prominent of the dichotomies that have fractured along with the brain/mind/body trichotomy. The world of models and theories has increasingly come to be dominated by complexity, non-linearity, fractals, chaos, and multi-logical systems. Hierarchies and dichotomies have been abandoned in favor of networks which are better suited to the complexities and connectivities we have begun to encounter in our search for causes and consequences across the sciences.

Postmodernists often seemed to be on the edge of or falling into the abyss of nihilism or the trap of naïve relativism. On the positive side, they provided insights on how to realistically revise our ideas about truth, objectivity, and science. Most importantly, they helped direct our attention to the social and cultural causes and contexts of our thoughts and behaviors. Network thinking has helped to transform the way we understand the brain. The networked social brain is the topic of this chapter. In mirroring complexity and non-linearity it will feedback and become prologue to Chap. 1. I begin with the story of how we arrived at current models of the networked brain.

The Social Intelligence Hypothesis

Our story begins with the social intelligence hypothesis. This hypothesis was designed to explain the size of human brains. The idea was that increasingly complex and dense social interactions were the driving forces behind the large brains of Homo sapiens. Around two million years ago the brain more than doubled in size reflecting the fact that humans were living in larger and larger and more and more complex groups. A larger brain reflected the need for a larger mental capacity to keep track of people and relationships in enlarging social groups. Then a second period of growth in the brain to its modern size occurred between 600,000 and 200,000 years ago. This period of growth was thought to be most likely related to cultural growth and especially to the evolution of language. Social intelligence was considered a critical factor in brain growth.

There are three possible relationships between brain size and social complexity. Either increasing social complexity led to larger and larger brains, or the evolution of larger and larger brains led to increasing social complexity. The more plausible hypothesis is that brain size and social complexity co-evolved.

The social intelligence hypothesis evolved into the social brain hypothesis. This development was based on research that showed that primates have unusually large brains for their body size among the vertebrates. The explanation takes off from the social intelligence hypothesis; larger brains evolved to cope with increasingly complex social systems. In fact, the general hypothesis, applied to all vertebrates, is that brain size is correlated with social complexity. It was initially hypothesized that primates evolved large brains to manage their unusually complex social systems. This hypothesis, originally applied in an evolutionary context to all vertebrate taxa, has been modified based on research that shows social brain differences in (using the classical distinctions) non-anthropoid primates, anthropoids, and other mammals. Brain size varies monotonically (see notes for definition) as a function of group size in primates. In other mammals and birds the relationship is qualitative. Larger brains are associated with differences in mating systems, in particular with pair-bonding. Researchers are uncertain about the reason for this association. We know two things: first, the anthropoid primates appear to have generalized pair-bonded mating to other non-reproductive relationships (e.g., "friendships"). What is at issue is why bonded relationships are so cognitively demanding that they require larger brains. This issue can be addressed by considering that social life in humans is not based on pair-bonding dyads but rather on social triads. Larger brains are associated with increasingly complex and dense social networks based on the triad as the fundamental unit of analysis. Keep in mind that I am not supporting a single causal pathway from brain size to social complexity or social complexity to brain size. Co-evolution is the most plausible way to view the relationship between brain and society.

The Social Brain Hypothesis

Leslie Brothers (1990) is widely credited with introducing the concept of the social brain to the neuroscience community. Brothers was well placed to make this introduction. She was not only a psychiatrist and neuroscientist but rare in her command of the sociological imagination. The model she introduced involved neural regions rather than the brain as a whole. The significance of her model is that it shifted the long-standing view of the brain as a biological entity essentially independent of social and cultural influences to a social entity. The traditional theories of brain and mind supported and were grounded in the myth of individualism. Brothers based her model on social brain research among non-human primates that came into prominence in the 1950s and thereafter. The increasing attention to the links between social complexity, social intelligence, and brain size fueled the social brain concept that Brothers finally crystallized in her 1990 paper.

How does a regional theory of the social brain work? It begins by asking the question: how do we infer the mental states of others? If we suppose that this ability is based on simulation and empathy, then (relying on a more or less traditional view of localization) we can identify this ability with the premotor cortex and the insula. If we assume a more organized theory of mind, then we are pointed to the medial prefrontal cortex and the temporo-parietal junction among other regions. The orbitofrontal cortex, a region in the frontal lobes, is associated with processing rewards. Lesions in this area are associated with severe disruptions in social behavior. At the same time cognition in other regions remains relatively intact.

The insula, located underneath the frontal cortex, represents bodily states such as pain and empathy; it is active when we become aware of the pain of others. The brain regions involved in social cognition are sensitive to context and their activation is modulated by social context and volitional regulation. All of this still leaves us with the mystery of explaining how inference works. The primates all show precursors to inference but it is most highly developed in humans. The explanation may lie in the "always, already, and everywhere social" theorem. To be social is to be connected, to evolve connected. We are cognitively coupled by virtue of being socially coupled to others. Compassion and empathy travel across that coupling. At this point some readers might be wondering about the research on mirror neurons. A mirror neuron fires when an animal acts and when it sees the same behavior in another. Such a mirroring system would seem to be a neurological corollary of the always, everywhere, and already social theorem. The conjecture is that the number of mirror neurons varies directly with the social complexity of social life in primates and perhaps in other animals (e.g. elephants). Brain activity consistent with that of mirror neurons has been found in humans. This is still a very controversial area and I cannot say much more with any confidence. However, some sort of mirror neuron system would seem to be a necessary piece of the social brain puzzle (for a review of the state of the art, see Ferrari and Rizzolatti 2014, 2015; Taylor 2016; Hyeonlin and Seung-Hwan 2018).

The first prominent appearance of the term "social brain" was in the book *The Social Brain* published in 1985 by the renowned psychologist Michael Gazzaniga. He made the bold move of linking biological and cultural forces in a serial causal nexus. He argued that causal forces determining our behavior and thought arise at the biological level and progress through the social and cultural levels. That viewpoint has prevailed into our own time among those neuroscientists who have actually taken social and cultural factors into account in their theories. When he comes around to explaining a specific complex human behavior, religious behavior, Gazzaniga falls back on a biological theory that is not informed by social theory even though he seems prepared to make this move.

Gazzaniga is not the first person to use the term "social brain." To my knowledge, the first use of the term occurs in a manuscript most likely from the 1950s written by the psychologist B.I. DeVore, "Primate Behavior and Social Evolution." The paper is cited by the anthropologist Clifford Geertz (1973: 68) as unpublished and undated. Geertz, as we will see, is one of the key starting points for a fully social theory of the brain.

The Brain in the Network

Recent developments at the nexus of the social-, life-, and neurosciences have witnessed the simultaneous invention of network models of systems and subsystems in the brain/mind/culture unit. The first hint of this development for me was the anthropologist Clifford Geertz's argument in 1973 for the synchronic emergence of an expanded forebrain among the primates, complex social organization, and at least among the post-Australopithecines tool savvy humans, institutional cultural patterns. This statement and his chapter a quarter century later (Geertz 2000) on brain/mind/culture, culture/mind/brain were the primary stimuli for the network model of the brain in culture and environment I developed with Sabrina Weiss about 20 years ago. The other important stimulus for our work was Mary Thomas Crane's *Shakespeare's Brain* published in 2000. Crane is an English professor who specializes in Renaissance literature and culture. She is representative of the trend toward interdisciplinary scholarship that marks the latter part of the twentieth century. She travels easily over the terrain of her specialty but just as easily in the lands of the cognitive sciences. The thesis she defends in her book is that biology engages culture and produces mind on the material site of the brain. This view articulates the architecture implied in Geertz's view of a network model of brain, mind, and culture.

The most recent iteration of my model appeared in 2017 and will be updated at the end of this chapter. Subsequently I discovered the works of neuroscientist Leah Krubitzer and sociologist Bernice Pescosolido. Krubitzer does not offer a specific model but rather a series of lessons learned in her research that have led her to realize that culture has played a key role in shaping our brains and behaviors. She stresses the significance of epigenetic research in redirecting us away from brains in vats to brains in contexts. The network concept shadows every one of the lessons she shares with us.

Pescosolido develops a sophisticated network model of the brain in context. She begins already knowing that the twentieth century ended with many scientists acknowledging the complex entanglements among genes, neurons, and behavior across different time scales. She concludes that we need a framework that links biological foundations, biological embedding, and social embeddedness. She calls that framework the Social Symbiome and develops it on the foundations of a Networks and Complex Systems science approach. The Social Symbiome has a kinship with the Connectome (see NIH's Human Connectome Project; and Hagmann 2005; Seung 2012). I will generalize the connectome concept in my model which already shows a kinship with that idea and the Social Symbiome.

Pescosolido's model illustrates the multiples hypothesis. In linking the brain to the social environment, she finds herself in the company of philosophers like Alva Noë and Andy Clark, the anthropologist Clifford Geertz, the biologist E.O. Wilson, the evolutionary neurobiologist Krubitzer, and me, a sociologist/anthropologist. Pescosolido's model is graphically different than mine but links "genes and proteins," "body," "self," "supports," "institutions," and "place" modeled respectively by molecular networked systems, biologically networked systems, individual systems ties, pathways to health/disease/health care, personal social network systems, organization based network systems, and geographical systems.

Reigning Myths About the Brain

Let's review some of the persistent myths about the brain that impact not only popular ideas about the brain but that continue to guide the research and thinking of many neuroscientists. Due to a misunderstanding and misinterpretation of the earliest split brain research, it became widely but mistakenly "known" or assumed that we operate within a relatively strict left brain/right brain paradigm. The left brain is alleged to support mathematical and logical skills; the right brain is supposedly the more artistic side. In fact while the brain does have two main segments connected by the corpus callosum, the two sides work wholistically and collaboratively. We think with our entire brain (and as we will see with our entire body in its social and environmental contexts). Left brain/right brain thinking is associated with localization theory, the theory that specific sections of the brain control specific functions such as language. Increasingly, neuroscientists have been forced to leave these traditional ideas behind and confront a new reality: culture plays an important role in determining what we do and think; and cognition or more generally mentality does not exist "in" the brain nor for that matter in the environment but in a systemic "in-between" space that links brain, mind, culture, and environment. It is important to recognize that this is a conclusion more readily reached if you approach the brain as a social scientist but it is a conclusion being forced upon neuroscientists by their own research.

In Chap. 4 I introduced the idea of social and neurological chaos. I am not using the term "chaos" in its everyday or common journalistic sense. Chaos is not a synonym for disorder. Over the last hundred years or so and especially during the past 50 years, we have learned a new vocabulary across the sciences, the arts, and the humanities. The key terms in this new vocabulary are "non-linear," "fractal," "complexity," "self-organization," "dynamical systems," and "chaos." This new vocabulary emerged as we learned the limits of traditional linear and simple billiard balls causal models; one of the key lessons of twentieth-century science and culture was that everything was more complex than we thought. One of the practical lessons of this new understanding is that we now recognize that working in a deliberate heavy handed way to create order in organizations, environments, and our lives is less useful than drawing on approaches that are more flexible and free-wheeling. This is a lesson the evolutionary process has already incorporated in a tinkering paradigm (Jacob 1977; Racine 2014). A short-hand management protocol would recommend a "more is less" approach, focus more on outcomes than on processes, give up the idea of a universal order, and recognize that there is order in disorder and patterns in randomness.

Classically, neuroscientists have sought the causal origins of human behavior at the level of individual neurons (the neuron doctrine), at the level of the neuronal organelle (e.g., the synapse or dendritic spine), or at the level of the neural network. These are micro-level versions of the macro-level idea of the "brain in a vat" and might be thought of as "neuron in a vat" models. In both cases, the brain or its neurons are isolated causal reservoirs that are hypothesized to cause our behaviors and thoughts. In general the neuron(s) in a vat model treat behavior in terms of a reflex model—a stimulus-response model. At least since the 1990s it has become apparent that the neuron(s) in a vat s-r model are not up to the task of explaining the complexities of human behavior. Newer approaches have drawn on self-organization theory and non-linear approaches in general including fractals and chaotic dynamics.

Non-linear dynamics has allowed neuroscientists to identify and study neuronal functioning at the level of the "cooperative neural mass." The implication of this line of research is that cortical functioning is internally self-organized. Perception, for example, is not a passive process in which the brain simply registers whatever stimulates the receptors. Perception begins within the brain in self-organized neural activity that lays the foundation for processing input. Internally generated chaotic dynamics is selective about what receptor activity to accept and process. To generalize, behavior is interactive; the brain in a sense reaches out toward input and gives it form through a self-organized process of patterning. The brain is the locus for this process which gives form and meaning to inputs. But once again I must stress that this brain work does not take place in isolation from body, culture, and environment.

Chaos, once thought to be neurologically pathological, turns out to be essential to the healthy functioning of the brain. What once was thought to be "noise" that needed to be filtered and eliminated turns out to be relevant behavioral signaling. Chaotic dynamics operates all across the brain at all levels in a network of functioning systems. It operates as the basal background state that keeps neurons exercised so they don't all die. This makes the stability of the brain independent to some extent of stimuli from the environment. But again, chaotic dynamics operates across the entire brain/body/culture/environment system.

Chaos is constitutive of complex networks and essential to the creation and stable circulation of information. In a nutshell, the brain does not just blindly process inputs. It is selective. It can be contrasted with "machines" which use periodic or steady-state dynamics. A chaotic system interacts with environmental input using an internally generated activity pattern.

Neuroscientists who have pioneered in and developed this model have taken important steps toward giving us a better understanding of how the brain works. However, things are more complicated even than this. My conjecture is that chaotic dynamics operate in a way that complicates tendencies toward hierarchical organization and promotes heterarchical organization. This applies not just to the brain but to the cell-to-gene-to neuron, to environment/society/culture system. That system is the unit across which chaotic dynamics operates. And it implies that the human umwelt is not unitary but varies as a function of culture. The umwelt is the world as it is experienced by specific organisms (see notes for the literature). Our views through our cultural lenses are as different from each other as the overarching human umwelt is different from the umwelt of other animals. This enhances the difficulties of cross-cultural communication beyond the centripetal force of compassion and the cultural construction of the "we" and the "they." Our differences are much more profound and resistant to the Other because they are rooted in an umwelt. Umwelts are culturally and environmentally patterned; that is, the differences that matter are the differences between human groups and cultures and between different animal species. Culture is a speciating mechanism; we can speak of cultural species in more or less the same way that we can speak of biological species.

It may have already occurred to some readers that the preceding evokes images of anarchy. And indeed we are now in the realm of anarchy. "Anarchy," like the term "chaos," as I use it here has nothing to do with popular views of bomb throwers, disorder, and the breakdown of the rule of law. Anarchism, as understood in the tradition that stems from Peter Kropotkin, is one of the sociological sciences and anarchy is a specific form of order, a specific way of organizing lives, organizations, and environments. I first used the phrase "the anarchy of the brain" in 2011 but more as a placeholder for an idea than a substantive idea. I now want to unpack what it means to say that the brain is organized according to the principles of anarchy.

It is difficult to discuss the chaotic dynamics of information theory and the brain without some technical details creeping in. We can adopt another perspective that can help explain the meaning and significance of a fine grained review of chaotic dynamics by looking at the relatively gross level of anarchic organization.

Students of the brain from inside and outside the neuroscience community have been prompted to introduce the concept of anarchy into their research for a number of reasons. The brain does not have a rigid structure. What I discussed above as chaotic dynamics is reflected in the constant structural modifications observed as the brain responds to internal, environmental, and social stimuli.

Perhaps the most important development in brain studies in the last 30 years is the recognition that the brain isn't "in a vat," it is *in* the world with the body; and it *has* a world and a body. Philosophers reached this conclusion a little earlier than the neuroscientists, but increasingly we have become aware of the arguments against the brain in a vat model. "I"—meaning the cells-environment and society unification (hereafter, I describe my model as "the unification model" or U-model)—am constantly processing the input I receive from my environment. We adapt to the world through a continuing plastic and analytical activity that characterizes all the systems and subsystems of the U-model. The plasticity of the brain is reflected in the plasticity of the body but also in the plasticity of all the other elements of the unification. This has to be the case because all systems and subsystems are non-unitarily isomorphic to one another. Plasticity has its limits in all systems; it inevitably encounters a fundamental recalcitrance in nature.

If we analytically distinguish the brain and the body, their adaptability is a function of their plasticity which at a more fine grained level reflects chaotic dynamics. The codes underlying the various levels of the U-model are not rigid. It is this feature of our species that explains our individual and collective abilities to adapt to the world, elaborate it, imagine it, and in a sense create it. However, the codes are constrained by natural parameters that represent the recalcitrance of the world. The mistake some scientists and others make is to assume that the lack of rigidity and the operation of chaotic dynamics is a basis for thought and action without limits.

We can say this in another way. Complex systems operate in accordance with precise rules but they are so numerous and varied that they cannot be captured in a few laws or equations. Hidden in the search for a Theory of

Everything or a single Equation of the Universe is the search for a God-surrogate. What we have in reality, speaking loosely, is a general uncertainty principle reflected in brain studies and in my U-model. It might be possible in principle to copy and reproduce the chemistry and structure of the brain but it would be impossible to copy and reproduce the electromagnetic content of the brain. We cannot hope to fill in all and perhaps any of the data (information packets; information quanta?) that represent our living thoughts and feelings. Chaotic dynamics and being in and having a world define a self—a social self to recall my earlier remarks—that cannot be cloned from DNA or discovered in the tissues of a dead brain. We are talking about open systems that are not deterministic but as parts of a sufficiently regulated cosmos always lawful. The genetic code organizes a framework within the brain and the body in the world and that framework gets its shape and content from living beings acting in that world. The locus of our "living thoughts and behaviors" is not the brain per se or the body per se; the locus is the brain/mind/body/culture system operating in the world, in an environment of interactions with others and the physical and natural world. If we could have started here, it would have been obvious why Einstein is not only not his brain; he is not even "Einstein." Einstein is only Einstein as a living being acting in the world through a defined period of time and space and moving in and through social networks.

If for analytical purposes we once again isolate the brain as a system, we notice that one of its key characteristics is the search for patterns. Once it recognizes a pattern it creates a kind of "information bowl" into which it throws any input that matches, more or less, that pattern. Chaotic dynamics keeps shaking up the bowls so patterns reign but do not rule despotically. This is crucial to creative thought. One of the most important behavioral causes and reflections of chaotic dynamics is humor. We can imagine it working in the following way. Humor shakes up tendencies toward rigidifying patterns of thinking and behavior. It tosses the information in a brain bowl into a mixer. The result is at least a temporary novelty in our experience that is "recorded" as the contents of the bowl return more or less to its original configuration. In a sufficiently creative situation of self and context, that configuration never in fact returns fully to any original configuration. There is a playfulness that many of the great scientists recognize to be of a piece with doing science. Einstein certainly had this quality. Richard Feynman put it this way: Now, in order to work hard on something, you have to get yourself believing that the answer's over there, so you'll dig hard there, right? So you temporarily prejudice or predispose yourself—but all the time, in the back of your mind, you're laughing (Feynman 1979/1999: 199).

Toward a Model of the Social Brain

My model is designed to graphically represent and expand Geertz's 1973 remarks on culture and the brain: the expanded primate forebrain, complex social organization, and institutional cultural patterns among the post-Australopithecines emerged and evolved more or less synchronously. This idea anticipates recent arguments for the coupled evolution of brain, mind, environment, and society.

This perspective recommends against treating biological, social, and cultural parameters as serially related in a causal nexus. Rather, these levels should be viewed as reciprocally intertwined and conjointly causal. The claim in a nutshell is that human behavioral repertoires emerge from the complex parallel and recursive interactions of molecules, cells, genes, neurons, neural nets, organs, biomes, the brain's various regions, the central nervous system, other elements of the body's systems and subsystems and our social interactions in their ecological and umwelt contexts. This implies that we need to re-think socialization.

Socialization is traditionally understood to operate on the person. My model re-imagines socialization as a process that simultaneously informs and variably integrates the biological self, the neurological self, and the social self to construct personality and character. In addition, each element in the model must be understood as a dialectical entity containing its own internal "seeds" of change, and as following a temporal dynamic that may be at different times synchronous or dyssynchronous relative to other elements. Each element is conceived as an information system with all systems multiply inter-linked by the circulation of information.

All systems, subsystems, and elements of the model are implicated in every thought and action but to different degrees and strengths at different times. They all have causal potential. In the case of a particular failure, a departure from the norm, a deviation in thought or action the dominant mechanism(s) in explaining the failure, departure, or deviation is(are) only a subset of the system or a particular configuration of elements. The entire network is technically involved but the peripheral causalities can—again to different degrees in different cases—be essentially ignored. They are in general "incidental" in the sense that the causal connection to the dominant mechanism(s) is weak or otherwise muted. We need to know what to focus on. The correct sub-network is the one that isolates the central relationships that allow us to understand what has causal priority among all the things going on; where should we intervene if we want to fix what's wrong? If something goes wrong with your car you don't anticipate rebuilding and reconfiguring the entire car.

Some recent findings on how the brain is structured will eventually have to be taken into account as my model evolves in relationship to research across the sciences. In trying to understand the complexity of the brain some scientists have tried to associate it with something they are familiar with. Scientists at Aberdeen University mapped the brain network to the universe. This allowed them to identify high dimensional objects that are the key to understanding structure and function. They used algebraic topology to model these structures in a virtual computer generated brain. The results were then tested on real brain tissue (Markram 2008; Reimann et al. 2017; Dean 2018).

They found that the brain reacts to a stimulus by building then razing a tower of multi-dimensional blocks on various geometrical scales from rods to planks to cubes (that is from one to three dimensions and on to higher dimensions). This progression is analogous to building a multi-dimensional sand castle (the "castle" is better described as "materializing") that then self-destructs. One of the structural implications of this viewpoint is the localization of memories in higher dimensional cavities. The dotted diagonals in my model represent a global basal chaotic function based on my earlier discussion of chaos and the brain. The following model was developed against the background of the research and theory on the social brain. This is less a formal model than it is a graphic representation of a new way of thinking about the brain in the world and in culture and society.

CODA

The research frontiers in the neurosciences are moving very rapidly and brain researchers are regularly announcing new and surprising discoveries. As a neuroscience watcher, I'm alert to the fact that developments even within the brains in a vat neuroist framework could radically alter our understanding of the social brain. It's hard to say in what ways or directions but there are still so many things we don't know about the brain per se that it is not unreasonable to expect changes that will impact my model coming from newly discovered features of the brain. I am confident that however radical these changes are they will not overturn the basic message of my model and social brain thinking: our interactions and environment impact the brain. Stay tuned.

A Transition to Issues of Practice and Clinical Perspectives

The story of Einstein's brain and Einstein the genius of geniuses has been viewed here through the lens of sociology and the social brain paradigm. In Chap. 6 I explore with the aid of two collaborators what the social brain paradigm means beyond the case of Einstein, genius, and creativity. The social brain paradigm does more than help us reveal the social and world realities beyond the conspiracy of mythologies that gives us Einstein as a grammatical illusion and his brain as a sacred relic. The social brain has implications for how we understand the brain in health and illness. In Chap. 6, we explore the implications of the social brain paradigm for the practice of identifying, preventing, and treating mental health issues.

Appendix 1: The Social Brain Model: *The Social Ecology of the Brain*

Figure 5.1 The original version of this model was designed with Sabrina Weiss. I have taken it through a number of revisions designed to keep pace with developments in neuroscience and in social neuroscience and neurosociology. It was designed to graphically represent and expand Clifford Geertz's argument for the synchronic emergence of brain and culture. In review, then: (1) biological, social, and cultural causal forces are reciprocally intertwined and conjointly causal; (2) human behavioral repertoires emerge from the complex parallel and recursive interactions of cells, genes, neurons, neural nets, organs, biomes, the brain and central nervous system, other elements of the body's systems and subsystems, and our social interactions in their ecological and umwelt contexts; (3) socialization is re-imagined as a process that simultaneously informs and variably integrates the biological self, the neurological self, and the social self to construct personality and character; (4) each element in the model is a dialectical entity containing its own internal "seeds" of change, and as following a temporal dynamic that may be at different times synchronous or dyssynchronous relative to other elements; (5) each element is conceived

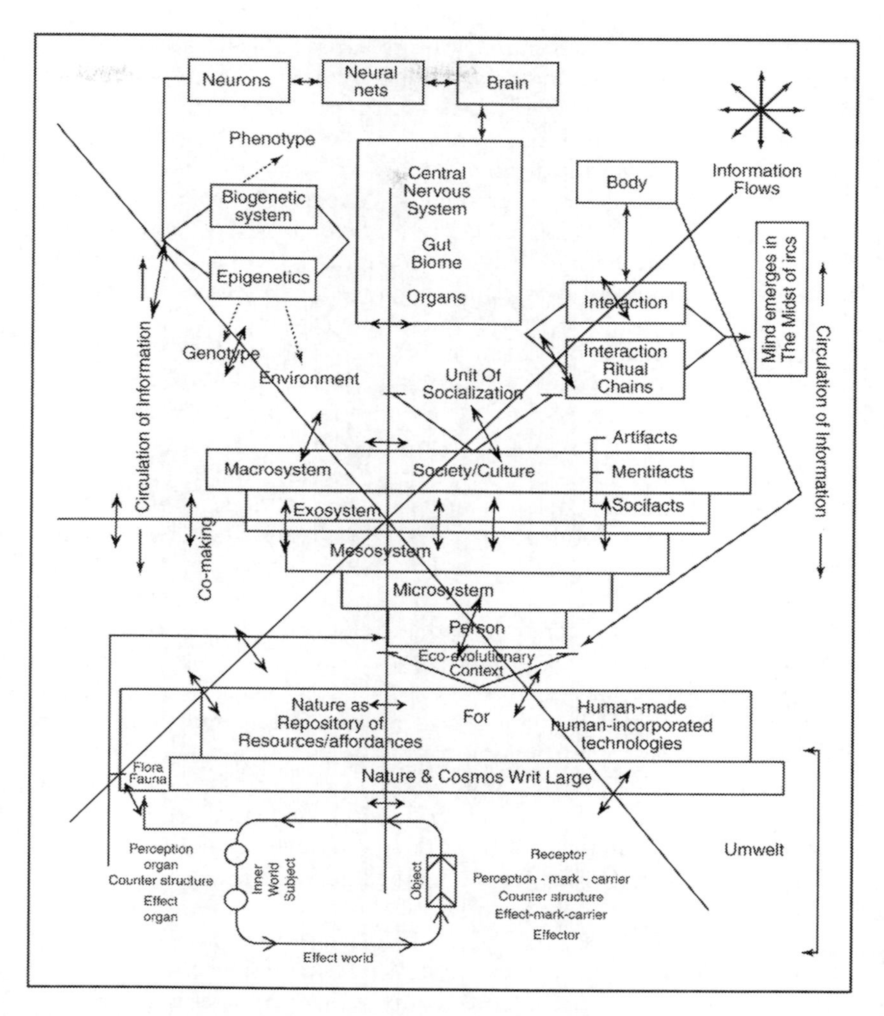

Fig. 5.1 The networked social brain

as an information system with all systems multiply inter-linked by the circulation of information; (6) the diagonals with double-headed arrows which crisscross the model map the chaotic dynamics and cooperative neural mass discussed by C.A. Skarda and W.J. Freeman (1987, 1990); (7) the unit model is activated in a triad of unit models and it is that triad that is the basic model of brain/mind/culture/world. This reflects the idea that the triad is the basic unit of social life (Restivo et al. 2014: 104n1); and (8) the diagram is the General Connectome. A connectome maps the elements and interconnections in a network. The term has been used specifically in connection with mapping the neural connections in the brain. Connectomes may range in scale from maps of parts of the nervous system to a map of all of the neural interactions in the brain. Partial connectomes have been constructed of the retina and primary visual cortex of the mouse. In line with these developments, my model represents the highest level of the connectome, a connectome of connectomes.

Based on the ideas introduced in the previous chapters I can now offer an initial concept formula for the probability of an "innovative thought." $iT_p = qc^2 \times K + G$, where qc^2 is the amount of cultural capital the person commands and K is a constant that represents the cultural context and network structure the person is embedded in; qc^2 because doubling the amount of cultural capital, for example, quadruples its impact factor. $K = C + N_t$. C = Cultural Context, an index that takes into account a variety of demographic, class, gender, and institutional diversity indicators; N = the density and diversity of the network structure of the society. G = the genius cluster quotient at time t. When considering the etiology of behaviors traditionally considered to be genetically grounded, it is now important to recognize that the brain, like humans, arrives on the evolutionary stage always, already, and everywhere, social. Therefore, what we have considered to be linearly transmitted genetic phenomena must now be viewed in the context of a brain that is at no stage of development separated from the social and cultural imperatives that form us. The very notions of "genes" and "genetic" must now be revised in the context of the social brain paradigm.

The next stage in this project is to embed the basic triad of the General Connectome in the nested networks of the social and cultural connectomes locally, regionally, and globally so that we now visualize a Global Connectome driven by the circulation of information across nested networks. On the rationale for a global connectome (my interpretation), see Khanna (2016) on "connectography."

Appendix 2: Connectomics

A connectome comprehensively maps neural connections in the brain. More broadly, a connectome maps all the neural connections in an organism's nervous system. Hagmann (2005) and Sporns et al. (2005) independently and simultaneously introduced the term "connectome," inspired by the efforts to construct a genome. Connectomics is the science of assembling and analyzing connectome data sets. Hagmann and Sporns discussed research strategies for developing comprehensive structural descriptions of the brain's networks, a dataset they called the "connectome." Such a connectome would help us understand the emergence of functional brain states from their structural substrate. Connectomics, the production and study of connectomes, can be applied at different scales from the full set of neurons and synapses in a part or all of an organism's nervous system to macro-level descriptions of the connections between all cortical and subcortical structures. The full connectome of the roundworm has been constructed along with partial connectomes of a mouse retina and primary visual cortex.

Appendix 3: Guides to the Technical Brain Terms Used in this Book

https://www.dana.org/brainglossary/
https://www.brainfacts.org/glossary
https://mayfieldclinic.com/pe-anatbrain.htm
https://www.google.com/search?q=parts+of+the+brain+and+their+functions+chart&safe=off&client=safari&sa=X&rls=en&biw=1218&bih=752&tbm=isch&source=iu&ictx=1&fir=nYBnlzaBiedEcM%253A%252CdE9NmhvLUJFzjM%252C_&vet=1&usg=AI4_-kTdU8tqEE14iJcfb1jdKW-J-0HWcg&ved=2ahUKEwiU3s709YTjAhUMZd8KHR-TDJcQ9QEwBXoECAcQDg#imgrc=_&vet=1

Appendix 4: Bibliographic Notes for Chapter 5

The Social Intelligence Hypothesis

Websites: for an overview of the social intelligence hypothesis and some initial references to evolution and the social brain see: https://www.sciencedirect.com/topics/psychology/social-intelligence-hypothesis; https://www.sciencedirect.com/science/article/pii/S1364661306003263: more recent examinations of the social intelligence hypothesis suggest the need for a broader theoretical framework that embraces "both inter-specific differences and similarities in cognition; ... how selection pressures that are associated with sociality interact with those that are imposed by non-social forms of environmental complexity, and how both types of functional demands interact with phylogenetic and developmental constraints" (Holekamp: https://doi.org/10.1016/j.tics.2006.11.003). For a critical evaluation of the hypothesis by Hemelrijk (2007) see: https://pdfs.semanticscholar.org/8df0/5c96641806c2bc45b220aeb792f563ee6473.pdf

Books, Articles, and Chapters

Whiten (2000), de Waal et al. (2003), Johnson-Ulrich (2017).
On the meaning of "monotonic": In mathematics, a monotonic function (or monotone function) is a function between ordered sets that preserves or reverses the given order. In calculus, a function defined as a subset of the real numbers with real values is called monotonic if and only if it is either entirely non-increasing or entirely non-decreasing. That is, a function that increases monotonically does not exclusively have to increase, it simply must not decrease.

The Social Brain Paradigm: Selected References

The literature on social cognition is relevant here but tends traditionally to be too embedded in the psychological-biological-neuroscience context which trumps the social and cultural contextual approaches. For an important exception, see Fiske and Taylor (2013); the authors are still under the influence of the idea that biology has causal priority over culture; but this is a serious effort to integrate emerging developments in social cognition with developments in social neuroscience, cultural psychology, and applied psychology. On the social brain per se, see Brüne et al. (2003), Dunbar et al. (2010); on social neuroscience, see Cacioppo et al. (2002), Schutt et al. (2015); on neurosociology, see Franks and Smith (1999), Franks and Turner (2013), Pickersgill and Keulen (2012), Barta (2014): this is really an anthropology of consciousness and extremely well done except for Bartra's effort to save free will.

On Issues Surrounding the Idea of the Split Brain

https://www.health.harvard.edu/blog/right-brainleft-brain-right-; https://brainconnection.brainhq.com/2001/06/26/roger-sperry-the-brains-inside-the-brain
https://www.inc.com/jessica-stillman/left-brained-v-right-brained-people-is-a-total-myt.html

The Chaotic and Anarchic Brain

Skarda and Freeman (1987, 1990), Duke and Pritchard (1991), Lehnertz et al. (2000), Soresi (2014), Zapporoli et al. (2015), Carhart-Harris and Friston (2019).

On the Concept of the Umwelt

See: Kull (1998), Sebeok (1976), Sebeok and Umlker-Sebeok (1978), and Uexküll, J.v. (1957, 2010/1934), and Uexküll, T.v. (1987).

References

Barta, R. 2014. *Anthropology of the Brain: Consciousness, Culture, and Free Will.* Cambridge: Cambridge University Press.

Brothers, L. 1990. The Social Brain: A Project for Integrating Primate Behavior and Neurophysiology in a New Domain. *Concepts in Neuroscience* I: 27–51.

Brüne, M., H. Ribbert, and W. Schiefenhövel, eds. 2003. *The Social Brain: Evolution and Pathology.* West Sussex: John Wiley & Sons.

Cacioppo, J.T., G.G. Berntson, R. Adolphs, et al., eds. 2002. *Foundations in Social Neuroscience.* Cambridge, MA: MIT Press.

Carhart-Harris, R.L., and J. Friston. 2019. REBUS and the Anarchic Brain: Toward a Unified Model of the Brain Action of Psychedelics. *Pharmacological Reviews* 71 (3): 316–344.

Dean, S. 2018. The Human Brain Can Create Structures in Up to 11 Dimensions. https://www.sciencealert.com/science-discovers-human-brain-works-up-to-11-dimensions.

Duke, D.W., and W.S. Pritchard, eds. 1991. *Measuring Chaos in the Brain.* London: World Scientific.

Dunbar, R., C. Gamble, and J.G. Owlett, eds. 2010. *Social Brain: Distributed Mind.* Oxford: Oxford University Press.

Ferrari, P.F., and G. Rizzolatti. 2014. Mirror Neuron Research: The Past and the Future. *Philosophical Transactions of the Royal Society B* 369 (1644). https://doi.org/10.1098/rstb.2013.0169.

———. 2015. *New Frontiers in Mirror Neurons Research.* Oxford: Oxford University Press.

Feynman, R. 1979. *The Smartest Man in the World.* OMNI interview; reprinted as pp. 189–204 in Richard Feynman (1999). *The Pleasure of Finding Things Out.* New York: Basic Books.

Fiske and Taylor. 2013. *Social Cognition from Brains to Culture.* London: Sage.

Franks, D.D. and T. Smith, eds. (1999), Mind, Brain, and Society: Towards a Neurosociology of Emotion, Vol. 5 of Social Perspectives on Emotion (Stamford, CT: JAI Press).

Franks, D.D., and J.H. Turner, eds. 2013. *Handbook of Neurosociology*. New York: Springer.

Geertz, C. 1973. *The Interpretation of Cultures*. New York: Basic Books.

———. 2000. *Available Light*. Princeton: Princeton University Press.

Hagmann, P. 2005. *From Diffusion MRI to Brain Connectomics Hampshire*. PhD Thesis. Lausanne: Ecole Polytechnique Fédérale de Lausanne.

Hyeonlin, J., and L. Seung-Hwan. 2018. From Neurons to Social Beings: Short Review of the Mirror Neuron System Research and its Socio-Psychological and Psychiatric Implications. *Clinical Psychopharmacological Neuroscience* 16 (1): 18–31.

Jacob, F. 1977. Evolution and Tinkering. *Science* 196 (4295): 1161–1166.

Johnson-Ulrich, L. 2017. The Social Intelligence Hypothesis. In *Encyclopedia of Evolutionary Psychological Science*, ed. T.K. Shackelford and V.A. Weekes-Shackelford, 1–7. New York: Springer.

Kull, K. 1998. On Semiosis, Umwelt, and Semiosphere. *Semiotica* 120 (3/4): 299–310.

Khanna, P. 2016. *Connectography: Mapping the Future of Global Civilization*. New York: Random House.

Lehnertz, K., C.E. Elger, J. Arnhold, and P. Grassberger, eds. 2000. *Chaos in Brain? Proceedings of the Workshop*. University of Bonn: Germany.

Markram, H. 2008. Fixing the Location and Dimensions of Functional Neocortical Columns. *Human Frontiers Science Program Journal* 2 (3): 132–135.

Pickersgill, M., and I.V. Keulen, eds. 2012. *Sociological Reflections on the Neurosciences*, Advances in Medical Sociology. Vol. 13. Bingley, UK: Emerald Group Publishing, Ltd.

Racine, V. 2014. Evolution and tinkering (1977), by Francois Jacob. Embryo Project Encyclopedia. https://embryo.asu.edu/pages/evolution-and-tinkering-1977-francois-jacob.

Reimann, M.W., M. Nolte, et al. 2017. Cliques of Neurons Bound into Cavities Provide a Missing Link Between Structure and Function. *Frontiers in Computational Neuroscience* 11 (48): 1–16.

Restivo, S., S. Weiss, and A. Stingl. 2014. *Worlds of ScienceCraft*. New York: Routledge.

Schutt, R.K., L. Seidman, and M.S. Keshavan, eds. 2015. *Social Neuroscience: Brain, Mind, and Society*. Cambridge, MA: Harvard University Press.

Sebeok, Thomas A. 1976. Foreword. In *Contributions to the Doctrine of Signs*, ed. T. Sebeok, x. Lanham, NH: University Press of America.

Sebeok, Thomas A., and D.J. Umlker-Sebeok. 1978. Linguistics: Speech Surrogates: Drum and Whistle Systems. *American Anthropologist* 80 (3): 712–713.

Seung, S. 2012. *Connectome: How the Brain's Wiring Makes Us Who We Are*. New York: Houghton Mifflin Harcourt.

Skarda, C.A., and W.J. Freeman. 1987. How Brains Make Chaos in Order to Make Sense of the World. *Behavioral and Brain Sciences* 10: 161–195.

———. 1990. Chaos and the New Science of the Brain. *Concepts in Neuroscience* 2: 275–285.

Soresi, E. 2014. *The Anarchic Brain*. Milano: Bookrepublic.

Sporns, O., G. Tononi, and R. Kötter. 2005. The Human Connectome: A Structural Description of the Human Brain. *PLoS Computational Biology* 1 (4): e42.

Taylor, J.M. 2016. Mirror Neurons After a Quarter Century: New Light, New Cracks. http://sitn.hms.harvard.edu/flash/2016/mirror-neurons-quarter-century-new-light-new-cracks/.

von Uexküll, Thure. 1987. The Sign Theory of Jakob von Uexküll. In *Classics of Semiotics*, ed. M. Krampen et al., 147–179. New York: Plenum.

———. 1957. A Stroll Through the Worlds of Animals and Men: A Picture Book of Invisible Worlds. In *Instinctive Behavior: The Development of a Modern Concept*, ed. Claire H. Schiller, 5–80. New York: International Universities Press.

von Uexküll, J. 2010/1934. *A Foray into the World of Animals and Humans With a Theory of Meaning*. Trans. J.D. O'Neil. Minneapolis: University of Minnesota Press.

de Waal, P.L. Tyack, et al., eds. 2003. *Animal Social Complexity: Intelligence, Culture and Individualized Societies*. Cambridge, MA: Harvard University Press.

Whiten, A. 2000: 185–196; discussion 196–201. *Social Complexity and Social Intelligence*. Basel, Switzerland: Novartis Foundation Symposium 233.

Zapporoli, L., M. Porta, and E. Paulesu. 2015. The Anarchic Brain in Action: The Contribution of Task-Based fMRI Studies to the Understanding of Gilles de la Tourette Syndrome. *Current Opinion in Neurology* 6: 604–611.

CHAPTER 6

The Social Brain: Implications for Therapeutic and Preventive Protocols in Psychiatry

Sal Restivo, Mario Incayawar, and Jenelle M. Clarke

Abstract The social brain paradigm does more than help us reveal the social and world realities beyond the conspiracy of mythologies that gives us Einstein as a grammatical illusion and his brain as a sacred relic. The social brain has implications for how we understand the brain in health and illness. The objective of this chapter is to explore the implications of social brain research and theory for the mental health professions. There is already a growing literature on the social brain and psychiatry. While this literature is an important rationale for our work, our approach is distinguished by a more strongly social science foundation for understanding the social brain. In Chap. 5, we saw how by building on the current body of knowledge about the social brain, we arrive at a new concept of the self and a new unit of socialization. Here we explore the implications of this re-envisioning of the social brain paradigm developed in Chap. 5 for clinical practice in psychiatry and prevention of mental illness.

Keywords Mental health • Mental illness • Psychiatry • Enriched environment • Clinical practice

S. Restivo, *Einstein's Brain*,
https://doi.org/10.1007/978-3-030-32918-1_6

Earlier chapters have demonstrated that it is widely assumed across the sciences and in the public arena that the neurosciences are the key to solving major mental and behavioral health problems. This assumption has driven important initiatives by Bush (1990) and Obama (2013) as well as brain initiatives in the EU (Brains in Dialogue workshops, 2009–2011), Japan (1997 Brain Science Institute), Saudi Arabia (The Brain Forum, 2013), and elsewhere since the turn of the century. The public support for these programs has been aided by colorful images created by neurotechnologies (e.g., fMRIs and PET scans) of "brains in action" doing things like playing chess or reading words. These images have been reinforced by dramatic headlines and book titles about "God in the brain" or "The moral brain." There is much to suggest that such headlines and images—often misleading and easily misconstrued and feeding neurohype-impact the way the public thinks about the brain (Konnikova 2014; and for an anthropological perspective on this phenomenon, see Dumit 2004; and see Restivo 2017: 133–135). The recent development within the neurosciences of the social brain paradigm urges that we reconsider this assumption. Additional support for the rationale guiding this chapter comes from calls for a new neuroscience-mental health agenda, multidisciplinary approaches to psychiatry, and a recent report on social experience and myelination published in *JAMA Psychiatry* (Holmes et al. 2014; Kinderman 2014).

The objective of this chapter is to explore the implications of social brain research and theory for clinical practice in psychiatry and the mental health professions. There is already a growing literature on the social brain and psychiatry (Kennedy and Adolphs 2012; Okasha 2009; Bakker et al. 2002). While this literature is an important rationale for our work, our approach is distinguished by a more strongly social science foundation for understanding the social brain (Bracken et al. 2012). In Chap. 5, we saw how by building on the current body of knowledge about the social brain, we arrive at a new concept of the self and a new unit of socialization. Here we explore the implications of this re-envisioning of the social brain paradigm for clinical practice in psychiatry and the treatment of mental illness in general.

Recall that in Chap. 5 we learned that Leslie Brothers (1990) introduced the concept of the social brain as a set of neural regions of the brain. Recent concepts are more wholistic but sustain a traditional conception of the brain as an independent unit of analysis (Dunbar 1998; Dunbar and Shultz 2007). The "nexus" version of the social brain treats biological, social, and environmental factors serially, with biological ones having causal priority. In the "network" model these factors are considered to be interrelated and to function conjointly with environmental and umwelt factors.

From the Networked Brain to the Psychiatric Clinic

Enriched Environment and Mental Health

On the foundations of social brain theory developed in Chap. 5, let us now explore the relationships between the "enriched environment theorem" and "Interaction ritual chain theory" in preventive and treatment protocols in psychiatry and mental health therapies. Based on the "enriched environment theorem," most health professional activities, occupations, and volunteer activities dealing with mental health can be defined as enriching environments. The implication of the social brain model and the enriched environment theorem is that the social sciences have greater relevance for psychiatry than is generally recognized. Notable social factors at work in the social brain include interaction ritual chains. This is a higher level of sociological generalization than that of social factors or social determinants of health which are more generally recognized in psychiatry, psychosocial interventions, and community mental health. These factors and determinants include gender, sex, race and ethnicity, socioeconomic status, household patterns, occupation, religious affiliation, labor markets, and social safety nets (McAlpine and Mechanic 2011; Abou-Saleh et al. 2011; and see Jacobs and Scheibel 1993; and Jacobs et al. 1993).

The enriched environment is more attuned with the recent realization that personal perception of relative (subjective) social status is better correlated with psychopathology than the traditional (objective) social status criteria such as educational level, income, and occupation (Scott et al. 2014). According to recent research, the personal everyday recurrent perception of financial strain is more relevant to mental health than objective financial wealth, as shown during the Great Recession (Wilkinson and Gerontol 2016). It could be argued that the subjective perception of social status better reflects the quality of the social environment enveloping the individual. In the last decade, a trend to take into account patients' very personal experiences, pharmacological treatment responses, and illness narratives has been emerging in medicine and psychiatry.

In certain pioneering clinical/academic centers, person-centered care, the practice of personalized medicine, and precision psychiatry are being explored. This medical model aims to manage patients' health and disease based on the person's specific characteristics such as weight and height, sex and gender, age, and genetic predispositions to disease and variations in responses to drugs, diet, and physical and social environments. This is a significant departure from common medical practices which are based on "standards of care." The new person-centered approach in medicine and psychiatry wants to be sensitive to patient's subjective, personal, and intimate perception of illness and disease. At this point, the new person-centered medical approach converges with the social brain paradigm and sets the stage for mutual cooperation and synergy between medicine and the social sciences. This clinical approach could lead health professionals to make better diagnoses, to provide finely tailored treatment plans and psychotherapy, and ultimately improve the quality of care.

Therapeutic Communities

Interaction ritual chain theory (IRCT; Collins 2004) is currently being used in a study of therapeutic community concepts and innovative intervention programs (Clarke and Waring 2018; Clarke 2017a, b; Clarke et al. 2016). Therapeutic communities consider all aspects of community life as potentially therapeutic (Jones 1976). IRCT is a theoretical framework developed by Randall Collins that allows for the systematic analysis of social encounters (Collins 2004, 2014; Summers-Effler 2004). There are, however, relatively few studies that explore the impact of interactions during informal periods. Solidarity and emotional rhythmic entrainment, the process by which individuals become in synch with one another, are crucial for establishing and maintaining inclusion and producing positive change outcomes. These interactions enable the acquisition of sophisticated socio-cognitive skills (Trevarthen 2000), which could improve community-based preventive interventions in psychopathology cases.

The social brain paradigm's potential explanatory power has already been explored in the general literature on psychiatry (Kandel 1998; Charlton 2003) and psychotherapy (Orlinsky and Howard 1986) and for specific problems such as autism spectrum disorder (Pelphrey et al. 2011). However, the social brain paradigm advocated for psychiatry and community mental health suffers from the same limitations as the social brain paradigm defended by classically oriented neuro- and social scientists. That version of the social brain paradigm is at its core the old independent free-standing brain that is the causal locus of our behaviors and our minds. At the same time, the paradigm is defended in the context of a call for a model that unifies biological, psychological, and social factors. That call and the realization of a social brain model that is fully sociologized have been answered by Restivo and Weiss (2016; see Chap. 5 this volume). The Restivo-Weiss model solves as a first approximation the variety of mind/brain, mind/body, and brain/body dilemmas and paradoxes that have plagued thinkers for millennia.

Interaction rituals are more than formal or ceremonial rituals. They are everyday social situations of "mutually focused attention and emotion" that produce feelings of belonging (Collins 2004). IRCT provides a mechanism for systematically analyzing the processes and outcomes of social encounters (Collins 2014; Hallett 2003; Summers-Effler 2002; Rossner 2013). The social brain model acknowledges that meaning about ourselves, others, and the social world occurs through interaction, thus linking interactions with neurological development and functioning (Mead 1934; Brothers 1997).

Current research shows how social interactions directly affect the structure and function of the brain and body. Stress that originates in maladaptive social interactions can significantly impact the intestinal microbiome and in turn modulate brain function, the immune system, and induce psychological changes and psychopathology (Ho and Ross 2017; Clarke et al. 2013). Moreover, within the social brain paradigm, it is recognized that individuals are intrinsically motivated to satisfy needs such as the need to belong and the need to be liked (Summers-Effler 2004). For instance, research conducted by Clarke (2015) demonstrates that when new client members enter a therapeutic community, a primary concern is often to be accepted by others in the group. One client describes this process: "You feel like going up and saying, 'Hi I'm Jessie, can we be friends?' So, it's intimidating and scary." Thus, during therapy, it is particularly important that individuals feel accepted in their therapeutic interactions in order to form healthy attachments. Belonging and solidarity foster feelings of safety, mutuality, and trust that make therapeutic work possible (Clarke and Waring 2018). The basis of the impressive effects of healthy/unhealthy social interactions seems to be their impact on genes, cells, organs, and the brain of the individual (Shonkoff and Phillips 2000). For instance, psychotherapy and particularly cognitive-behavioral therapy seems to induce anatomical and functional changes in the brain in the same way as psychopharmacological treatment does. Similar changes of the caudate nucleus in the brain have been found after either interpersonal psychotherapy or the administration of a selective serotonin re-uptake inhibitor antidepressant paroxetine (Brody et al. 2001).

Social needs are met through interaction rituals in everyday encounters with others that are more or less simultaneously encoded in neural and other strata as prescribed in the U-model (Chap. 5). Establishing feelings of belonging and acceptance is essential for tolerating negative emotional experiences that occur throughout life. However, when these needs are not met, individuals may develop negative coping strategies that may result in mental distress (Summers-Effler 2004a).

The encoding process occurs via "sensory appraisal" of a mechanism called emotional energy, or EE (Collins 2004; Summers-Effler 2004). EE is a concept that describes all aspects of the social mood of a group, including its rhythmic expression. EE is generated within small and large groups, such as smoking breaks or sporting events, lingering within individuals after the group has dispersed and recalled through memory (Jones 1976). During memory recollection, subsets of neurons that were active during the encounter are reactivated, which creates an internal representation of the situation that is reflected in the body (Brothers 1997). For example, embarrassment during a social encounter may cause flushing in the face. During successful rituals, positive EE produces individual feelings of confidence and enthusiasm, as well as group feelings of belonging. Individuals are motivated to maximize EE, or at worse, to minimize EE loss (Collins 2004; Summers-Effler 2002). Moreover, rituals do not occur in isolation but are linked together in a "chain" of interactions over time (Collins 2004, 2014). Through these chains of interactions, individuals form expectations of social encounters. Thus the body, brain, and environment work together to create patterns and expectations of interpersonal interactions.

According to IRCT, as individuals interact, they become mutually and rhythmically entrained. This rhythmic entrainment builds alliance and solidarity between individuals that is necessary for healthy relationships. In other words, formation of attachments occurs through entrained, rhythmic interactions. Entrainment is physically experienced in the synchronization of speech and bodily movements, and emotionally felt when one is understood and validated. It is particularly important to establish entrainment during the expression of negative emotions. For instance, in Clarke's (Clarke et al. 2017) research, during informal time on a Saturday evening, a client member became very upset about her relationship with her mother and her disordered eating and admitted to the others that she had urges to self-harm. Other client members offered her tissues, shared their own experiences of difficult relationships, and thereby validated her distress. The client's distress visibly decreased and the community discussed alternative ways this client could manage her negative emotions without self-harming. Thus, establishing and maintaining entrainment is essential for generating solidarity and effective therapeutic relationships between clients and staff members.

The use of the concept of emotional energy affecting the person and a group of people could be useful in the implementation of primary prevention programs in psychopathology. It would be clearly an advancement of public mental health programs organized by private institutions, non-profit organizations, and the government. Usually primary prevention programs of psychopathology takes into account the social network factor in a very simplistic way. These community mental health programs operationalize the social network as the need to increase the number of friends, from zero to about three or five. The success of a program is measured by the demonstration that on average the number of friends of the target community has increased after a community intervention. Conventional primary prevention in psychopathology programs also consider interventions on perceived stress, biological vulnerabilities, social and economic hardships, as well as increasing coping skills, social networks, and sense of self-efficacy (Kessler and Goldston 1986; Joffe et al. 1984).

Entrainment in IRCTs is similar to Trevarthen's (2000; Trevarthen and Aitken 2001) model of neurocognitive development, which suggests that psychological and biological processes are fully embedded within a social and cultural context. Elements of IRCT thus parallel Trevarthen's analysis of the neuropsychological strata underlying parent and child interaction. This is illustrated in his description of successful and unsuccessful rhythmic entrainment in social groups. Therefore, healthy social environments and good parent/child interaction are essential for the development of the healthy social brain. However, highly disruptive and complex traumatic experiences in early life, often within maladaptive social environments, impact the acquisition of the sophisticated socio-cognitive skills necessary for healthy emotional and relational functioning in adulthood (Burns 2006; Roth and Fongay 2005; Summers-Effler 2002; Wurtele 1998). The impact of these experiences is often expressed through emotional, psychological, and behavioral difficulties such as dissociation, self-harm, and interpersonal conflict.

For individuals with a history of negative interactions involving abuse, trauma, and neglect, interaction rituals that generate positive EE can have a positive impact upon the social brain. Conflict between one's negative expectations of social situations and the actual encounter can prompt reflexivity and a conscious decision to update expectations based upon new meaning (Damasio 1994). In other words, healthy interactions directly challenge the expectation that all social encounters produce negative emotion and can change the brain's response to social stimuli. If, on the other hand, one's social expectations are not reinforced over time, these expectations break down and slowly shift (Summers-Effler 2004). In both scenarios, neurons and other elements in the social ecology of the brain become encoded with different emotional meaning.

This process of changing social expectations can be seen in planned social environments such as therapeutic communities (TCs). As noted above, TCs consider all interactions as potentially therapeutic. They are a psychosocial treatment modality that aims to provide a safe setting whereby troubling relational patterns and difficulties can be explored. One of the core aims of TCs is to help clients replace dysfunctional interpersonal styles with healthier ways of relating (Haight 2013). Within TCs, all aspects of community life, particularly everyday social encounters, are opportunities to change social expectations of interactions (Jones 1976). These everyday situations, or interaction rituals, include smoking breaks, meal times, and community meetings. It is often in these settings that rhythmic entrainment and positive EE are enacted in enhanced ways. Keep in mind that a base level of rhythmic entrainment is an automatic feature of any interaction (based on the "everywhere, always, and already social" theorem).

Crucial to the success of social environments is the emotional tone and rhythm of interaction rituals (Collins 2004). Emotions play a key role during social interactions as they provide the musical tone and pace of a ritual's rhythm, and provide individuals' expectations of future interactions, motivation for action and relational bonds. Thus, it is emotions within interaction that give rise to entrained rhythm. In Clarke's (Clarke and Waring 2018) work, for example, a client with disordered eating became quietly upset while struggling to finish her food at a mealtime. Another client at the dining room table immediately noticed something was wrong and, without hesitating, gently invited the other client to sit and talk with her while she finished her food. Emotional entrainment is therefore particularly important in developing and maintaining therapeutic alliance (Ackerman and Hilsenroth 2003). Establishing entrainment beyond the base level is essential and linked with positive emotional energy during therapy.

Maintaining positive emotional energy in social environments, which is necessary for producing change, is challenging. Rituals are not fixed or static and their level of stability is in part dependent on the participants engaged in the interactions. Variations in emotional tone and rhythm can change within a single ritual and impact future and surrounding rituals

(Hallett 2003). This emotional movement builds into three oscillations of rhythm within a social environment: interaction, person, and community. Changes in the person, such as fewer episodes of disassociation, have an impact upon the group as a whole. For instance, in a TC, if the community has more individuals who refrain from using self-harming behaviors, the community will feel stable and resilient. Conversely, if members are consistently using self-harming behaviors over a period of time, the emotional tone of the community will eventually change to match the overriding feelings of anxiety, mistrust, and frustration that arise from members and their interactions. These waves are depicted in Figs. 6.1, 6.2, 6.3, and 6.4:

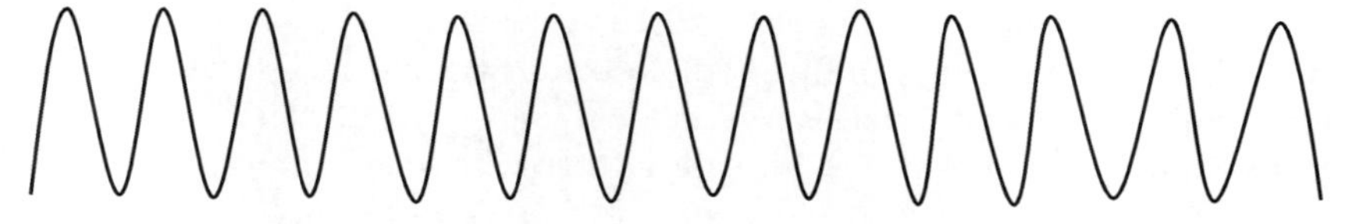

Fig. 6.1 Interaction rhythm wave

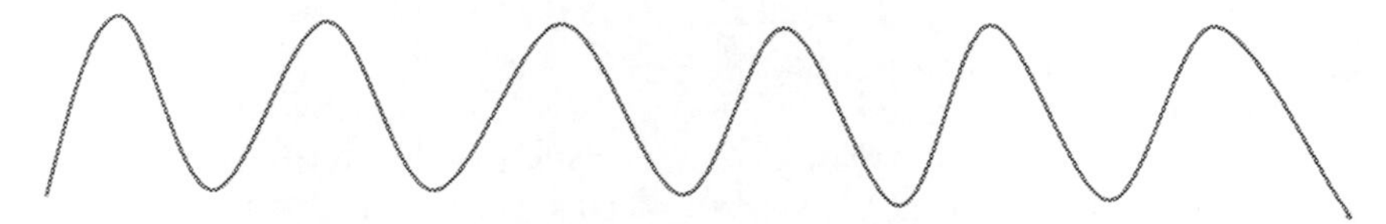

Fig. 6.2 Individual(s) rhythm wave

Fig. 6.3 Community or group rhythm wave

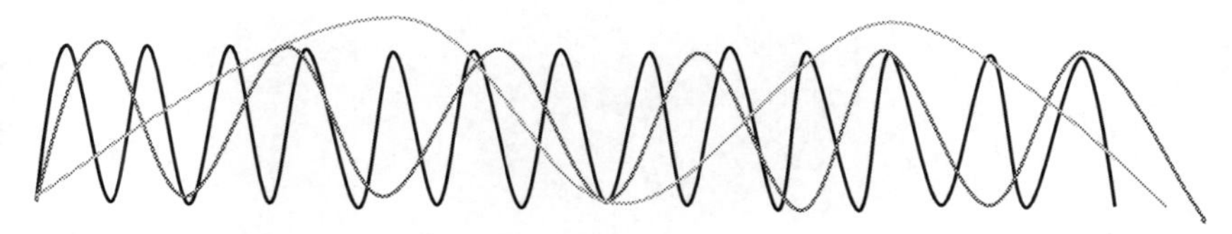

Fig. 6.4 Combined rhythm waves

These differing waves of rhythm occur simultaneously within a group and together produce emotional energy (Fig. 6.4). The quality of social interactions will directly influence the emotional mood of individuals and the dynamics within the social environment (Hallett 2003). Collins emphasizes that positive EE is made up of feelings such as enthusiasm and confidence, which he calls "higher-order social emotions" (Collins 2014). Over time, subtle changes in interactions can therefore generate emotional and rhythmic changes within individuals by changing their encoded neurological patterns of social relating.

This position does not help us fully understand how therapeutic change can occur, particularly when the therapeutic process often generates negative emotion within the social environment due to the distressing nature of the issues confronted. Research within TCs suggests that everyday community rituals can transform negative EE, beliefs, expectations, and behaviors into positive EE through social interaction. For example, emergency meetings, whereby the community as a whole addresses a member's risky behavior (e.g., suicide attempt), are often distressing for staff and clients, but ultimately can produce feelings of belonging and hope in individual members if solidarity and entrainment as a group are maintained.

Interaction rituals embedded within a neurocognitive framework provide a multi-layered explanation of the successful interpersonal therapeutic process. Patterns for social relating that are neurologically encoded are learned and re-learned through social interactions. Where individuals develop unhealthy forms of interpersonal relating, particularly as a result of troubling life experiences, social environments such as TCs can enable individuals to change their expectations of social encounters. Significantly, negative emotions that are present within groups can be transformed into positive EE if solidarity and entrainment are maintained. Therefore, solidarity and entrainment within everyday interaction rituals are crucial for social environments within mental health. What we have here are empirical demonstrations of enriched environments which research suggests have corollary neurological consequences.

Conclusion

The social brain paradigm has the potential to effectively bridge the social sciences to the medical, psychiatric, and mental health world. The social brain concept has more explanatory power than the usual biopsychosocial or social determinants of health models. In practice, it is going to nurture innovations and reform in clinical medicine and psychiatry to the benefit of community mental health interventions. Our understanding of the enriched environment allows us to see more easily an environmentally (social and physical) embedded brain, highly responsive to positive or negative changes in persons' surroundings. With this understanding mental health practitioners, community mental health workers, and mental health policy makers will be better equipped to develop a person-centered psychiatric care protocol. In the same way, the new notion of interaction rituals developed by the sociological sciences will allow better understanding of how we interact with others in everyday life. As a consequence, the psychiatric clinician will prepare more integrative holistic treatment plans and the mental health worker could design fine-tailored community interventions and prevention programs.

The social brain paradigm now needs to be tested and retested in a rigorous way. Hard evidence from clinical and public health programs based on the social brain paradigm is now needed. Ultimately, this effort will lead to enriched person-centered, socially and culturally sensitive health services, improved mental health policies, and a higher overall quality of care.

References

Abou-Saleh, M.T., C.L.E. Katona, and A. Kumar, eds. 2011. *Principles and Practice of Geriatric Psychiatry*. 3rd ed. New York: John Wiley & Sons.

Ackerman, S., and M.J. Hilsenroth. 2003. A Review of Therapist Characteristics and Techniques Positively Impacting the Therapeutic Alliance. *Clinical Psychology Review* 23 (1): 1–33.

Bakker, C., R. Gardner Jr., V. Koliatses, et al. 2002. The Social Brain: A Unifying Foundation for Psychiatry. *Letters, Academic Psychiatry* 6 (3): 219.

Bracken, P., P. Thomas, S. Timimi, et al. 2012. Psychiatry Beyond the Current Paradigm. *The British Journal of Psychiatry* 201: 430–434.

Brody, A.L., et al. 2001. Regional Brain Metabolic Changes in Patients With Major Depression Treated With Either Paroxetine or Interpersonal Therapy: Preliminary Findings. *Archives of General Psychiatry* 58 (7): 631–640.

Brothers, L. 1990. The Social Brain: A Project for Integrating Primate Behavior and Neurophysiology in a New Domain. *Concepts in Neuroscience* I: 27–51.
———. 1997. *Friday's Footprint: How Society Shapes the Human Mind*. New York: Oxford University Press.
Burns, J. 2006. The Social Brain Hypothesis of Schizophrenia. *World Psychiatry* 5 (2): 77–81.
Charlton, B.G. 2003. Theory of Mind Delusions and Bizarre Delusions in an Evolutionary Perspective: Psychiatry and the Social Brain. In *The Social Brain: Evolution and Pathology*, ed. M. Brüne, H. Ribbert, and W. Schiefenhövel, 315–338. New York: John Wiley & Sons.
Clarke, G., et al. 2013. The Microbiome-Gut-Brain Axis During Early Life Regulates the Hippocampal Serotonergic System in a Sex-Dependent Manner. *Molecular Psychiatry* 18 (6): 666–673.
Clarke, J.M. 2015. *Where the Change Is: Everyday Interaction Rituals of Therapeutic Communities*. Doctoral Dissertation. Nottingham: University of Nottingham.
———. 2017a. The Case for "Fluid" Hierarchies in Therapeutic Communities. *Therapeutic Communities: The International Journal of Therapeutic Communities* 38 (4): 207–216.
———. 2017b. The Role of Everyday Interaction Rituals within Therapeutic Communities. In *Mental Health: Uncertainty and Inevitability: Rejuvenating the Relationship Between Social Science and Psychiatry*, ed. H. Middleton and M. Jordan, 47–72. London: Palgrave Macmillan.
Clarke, J.M., and J. Waring. 2018. The Transformative Role of Interaction Rituals Within Therapeutic Communities. *Sociology of Health and Illness* 40 (8): 1277–1293.
Clarke, S., G. Winship, J. Clarke, and N. Manning. 2017. Therapeutic Communities. In *Psychiatric and Mental Health Nursing: The Craft of Caring*, ed. M. Chambers, 631–640. Abingdon, UK: Routledge.
Clarke, S.P., J.M. Clarke, R. Brown, and H. Middleton. 2016. Hurting and Healing in Therapeutic Environments: How Can We Understand the Role of the Relational Context? *European Journal of Psychotherapy & Counselling* 18 (4): 384–400.
Clarke, S., G. Winship, J. Clarke, and N. Manning. 2017. Therapeutic Communities. In *Psychiatric and Mental Health Nursing: The Craft of Caring*, ed. M. Chambers, 631–640. Abingdon: Routledge.
Collins, R. 2004. *Interaction Ritual Chains*. Princeton: Princeton University Press.
———. 2014. Interaction Ritual Chains and Collective Effervescence. In *Collective Emotions*, ed. C. von Scheve and M. Salmela, 299–311. Oxford: Oxford University Press.
Damasio, A. 1994. *Descartes' Error*. New York: G.P. Putnam.
Duke, D.W., and W.S. Pritchard, eds. 1991. *Measuring Chaos in the Human Brain*. New York: World Publishing.

Dunbar, R. 1998. The Social Brain Hypothesis. *Evolutionary Anthropology* 1 (8): 184–190.

Dunbar, R., and S. Shultz. 2007. Evolution in the Social Brain. *Science* 317: 1344–1351.

Haight, R. 2013. The Quintessence of a Therapeutic Environment. *The International Journal of Therapeutic Communities* 34 (1): 6–15.

Hallett, T. 2003. Emotional Feedback and Amplification in Social Interaction. *The Sociological Quarterly* 44 (4): 705–726.

Ho, P., and D.A. Ross. 2017. More Than a Gut Feeling: The Implications of the Gut Microbiota in Psychiatry. *Biological Psychiatry* 81 (5): e35–e37.

Holmes, E.A., M.G. Craske, and A.M. Graybiel. 2014. A Call for Mental Health Science. *Nature* 511: 287–289.

Jacobs, B., and A.B. Scheibel. 1993. A Quantitative Dendritic Analysis of Wernicke's Area in Humans. I. Lifespan Changes. *Journal of Comparative Neurology* 327: 83–96.

Jacobs, B., M. Schall, and A.B. Scheibel. 1993. A Quantitative Dendritic Analysis of Wernicke's Area in Humans. II. Hemispheric and Environmental Factors. *The Journal of Comparative Neurology* 327: 97–111.

Joffe, J.M., et al. 1984. *Readings in Primary Prevention of Psychopathology: Basic Concepts*. Lebanon, NH: University Press of New England).

Jones, M. 1976. *Maturation of the Therapeutic Community: An Organic Approach to Health and Mental Health*. New York: Human Sciences Press.

Kandel, E.R. 1998. A New Intellectual Framework for Psychiatry. *American Journal of Psychiatry* 155: 457–469.

Kennedy, D.P., and R. Adolphs. 2012. The Social Brain in Psychiatric and Neurological Disorders. *Trends in Cognitive Sciences* 11: 559–572.

Kessler, M., and S.E. Goldston. 1986. *A Decade of Progress in Primary Prevention*. Lebanon, NH: University Press of New England.

Konnikova, M. 2014. How Headlines Change the Way We Think. *The New Yorker*. https://www.newyorker.com/science/maria-konnikova/headlines-change-way-think.

Kinderman, P. 2014. *Prescription for Psychiatry: Why We Need a Whole New Approach to Mental Health and Wellbeing*. New York: Palgrave Macmillan.

McAlpine, D.D., and D. Mechanic. 2011. The Influence of Social Factors on Mental Health. In *Principles and Practice of Geriatric Psychiatry*, ed. M.T. Abo-Saleh, C.L.E. Katona, and A. Kumar, 3rd ed., 97–102. New York: John Wiley & Sons.

Mead, G.H. 1934. *Mind, Self, & Society*. Vol. 1. Chicago: University of Chicago Press.

Okasha, A. 2009. The Social Brain: A New Perspective. *Current Psychiatry* 16 (1): 1–6.

Orlinsky, D.E., and K.J. Howard. 1986. Process and Outcome in Psychotherapy. In *Handbook of Psychotherapy and Behavior Change*, ed. S. Garfield and A. Bergin, 3rd. ed., 311–381. New York: John Wiley & Sons.

Pelphrey, K.A., C.M. Hudac, et al. 2011. Research Review: Constraining Heterogeneity: The Social Brain and its Development in Autism Spectrum Disorder. *Journal of Child Psychology and Psychiatry* 52 (60): 631–644.

Restivo, S., and S. Weiss. 2016. The Social Ecology of Brain and Mind. In *Worlds of ScienceCraft*, ed. S. Restivo, S. Weiss, and A. Stingl, 37–70. New York: Routledge.

Restivo, S. 2017. *Sociology, Science, and the End of Philosophy: How Society Shapes Brains, Gods, Maths, and Logics*. New York: Palgrave Macmillan.

Rossner, M. 2013. *Just Emotions: Rituals of Restorative Justice*. Oxford: Oxford University Press).

Roth, A., and P. Fongay. 2005. *What Works for Whom? A Critical Review of Psychotherapy Research*. New York: The Guilford Press.

Scott, K.M., A.O. Al-Hamzawi, L.H. Andrade, et al. 2014. Associations Between Subjective Social Status and DSM-IV Mental Disorders: Results from the World Mental Health Surveys. *JAMA Psychiatry* 71 (12): 1400–1408.

Shonkoff, J.P., and D.A. Phillips, eds. 2000. *From Neurons to Neighborhoods: The Science of Early Childhood Development*. Washington, DC: The National Academies Press.

Summers-Effler, E. 2002. The Micro Potential for Social Change: Emotion, Consciousness, and Social Movement Formation. *Sociological Theory* 20 (1): 41–60.

Summers-Effler, E. 2004a. A Theory of the Self, Emotion, and Culture. Theory and Research on Human Emotions. In *Theory and Research on Human Emotions*, ed. J. Turner, vol. 21, 273–308. Bingley, West Yorkshire, UK: Emerald Group Publishing, Ltd.

———. 2004b. Defensive Strategies: The Formation and Social Implications of Patterned Self-destructive Behavior. *Advances in Group Processes* 21: 309–325.

Trevarthen, C. 2000. Musicality and the Intrinsic Motive Pulse: Evidence from Human Psychology and Infant Communication. *Musicae Scientiae* 3 (1): 155–215.

Trevarthen, C., and K.J. Aitken. 2001. Infant Intersubjectivity: Research, Theory and Clinical Applications. *Journal of Child Psychology and Psychiatry* 42 (1): 3–48.

Wilkinson, L.R., and B. Gerontol. 2016. Financial Strain and Mental Health Among Older Adults During the Great Recession. *Psychological Sciences and Social Sciences* 71 (4): 745–754.

Wurtele, S.K. 1998. Victims of Child Mistreatment. In *Comprehensive Clinical Psychology*, ed. A.S. Bellack and M. Hersen, vol. 9, 341–358. New York: Pergamon.

Sal Restivo is a sociologist/anthropologistQ1 who has held professorships and endowed chairs at universities in the United States, Europe, and China. He is a former president of the Society for Social Studies of Science and author most recently of *Sociology, Science, and the End of Philosophy: How Society Shapes Brains, Gods, Maths, and Logics* (Palgrave Macmillan, 2017).

Mario Incayawar is a physician-scientist, educator, and entrepreneur who specializes in psychiatry and pain medicine. He is the director of the Runajambi Institute, Montreal, Canada. His specialties include the bio-psychosocial dimensions of mental disorders, chronic pain and analgesia, pharmacogenetics of psychotropic drugs, drugs repurposing in psychiatry, precision medicine, and deep medicine. He is the author of many publications in these fields including *Culture, Brain, and Analgesia: Understanding and Managing Pain in Diverse Populations*, co-edited with K.H. Todd (2012); *Overlapping Pain and Psychiatric Syndromes: Global Perspectives* (co-edited with M.R. Clark and S. Maldonado-Bouchard) is forthcoming from Oxford University Press.

Jenelle M. Clarke University of Birmingham, holds a Ph.D. in Sociology and Social Policy from the University of Nottingham, UK. Drawing on interaction ritual theory, she has studied everyday social interaction in the setting of a therapeutic community. She specializes in social studies of healthcare policy and therapies. Her interests include microsociology, social work and mental health, ethics, and the sociology of communities. Her articles on therapeutic communities have appeared in journals such as the *European Journal of Psychotherapy & Counselling* and the *International Journal of Therapeutic Communities*.

Postscript

At this point, given all that has gone before, the reader should recognize that the author of this book like its subject is a grammatical illusion. "Sal Restivo" stands for a social network, rests inside a nested network of networks, is a voice box for these networks, stands on their shoulders, and is part of a creative cluster that is defined by its challenges to the illusions of the free willing unimpeded agential self and the brain as a free-standing font of all of our behaviors and thoughts. Social networks are reading this product of a social network. I stand amid adumbrationists, anticipators, independent inventors, precursors, and known and unknown companions in a collective effort. Some of these I knew of beforehand, some I discovered after my project was underway. This book self-confirms and self-exemplifies its theses.

Finally, errors of fact, commission, and omission will inevitably show up in this book, given its broad interdisciplinary approach. None of these will in the end change the basic facts of the mind and brain as social constructions. Social construction should not be viewed narrowly and as reductionist. Social construction arises at the intersection of social, biological, physical, and chemical processes. I have tried many times to correct mistakes about what "social construction" means. Briefly, it means that we can only invent and discover by working with others in social contexts and in generational networks that persist over time. Sometimes links are broken; but for science to progress, those links have to be reconnected—across cultures, across historical eras, and across ideological fractures. (On the

S. Restivo, *Einstein's Brain*,
https://doi.org/10.1007/978-3-030-32918-1

meaning of social construction, see Restivo and Croissant 2007; Restivo 2008). Social constructionism is the fundamental theorem of sociology and not some vague philosophical idea.

What can we expect concerning the evolution of Einstein's contributions? The historical record suggests they will not last forever. Every scientific discovery, fact, or theory closes off a line of inquiry. Closures vary by degree. The fact that the earth is not flat represents the highest degree of closure. It will not be resurrected due to new findings as science progresses. Our ideas about gravity, Newtonian and Einsteinian are closed to a lesser degree. Notice this: Newton closes a chapter in the history of theories about gravity; but the way he closes it in the context of the way science unfolds or progresses doesn't prevent Einstein. The Einstein closure does not eliminate the possibility of new theories of gravity and the cosmos. There have been critics reasonable and unreasonable since the publication of his papers on special and general relativity. For an introduction to these criticisms and calls for a post-Einsteinian physics, see the following websites. I begin with some of the less credible proposals (note in particular Al Kelly 2005):

https://www.google.com/search?clinet=safari&rls=en&q=Challenging+Einstein&ie=UTF-&&oe=UTF-8;
https://www.google.com/search?safe=off&client=safari&rls=en&ei=41FHXenoKqSt5wKo4bqABQ&q=Challenging+Einstein%2BA.+Kelly&oq=Challenging+Einstein%2BA.+Kelly&gs_l=psy-ab.3...82513.85496..86002...0.0..0.131.761.8j1......0....1..gws-wiz.......35i304i39j0i22i30j33i22i29i30j33i299j33i160.D-PgJJXH7Jc&ved=0ahUKEwip5qKKmOrjAhWk1lkKHaiwDlAQ4dUDCAo&uact=5

For more credible intimations of post-Einsteinian physics, see:

https://www.nbcnews.com/mach/science/einstein-showed-newton-was-wrong-about-gravity-now-scientists-are-ncna1038671: "We now have the technological capacity to test gravitational theories in ways we've never been able to before," study co-author Jessica Lu, an astrophysicist at the University of California, Berkeley, said. "Einstein's theory of gravity is definitely in our crosshairs." We may be closer than ever to witnessing the eclipse of Einstein's general theory; and see: https://www.newscientist.com/round-up/challenging-einstein/. Every scientific story implies the same ending: Stay tuned.

References

Restivo, S. 2008. Society, Social Construction, and the Sociological Imagination. *Constructivist Foundations* 3 (2): 94–96.

Restivo, S., and J. Croissant. 2007. Social Constructionism in Science and Technology Studies. In *Handbook of Constructionist Research*, ed. J. Holstein and J. Gubrium, 213–229. New York: Guilford Press.

References

Abou-Saleh, M.T., C.L.E. Katona, and A. Kumar, eds. 2011. *Principles and Practice of Geriatric Psychiatry*. 3rd ed. New York: John Wiley & Sons.

Abraham, C. 2002. *Possessing Genius: The Bizarre Odyssey of Einstein's Brain*. New York: Vintage.

Aaman, K., and Karin Knorr-Cetina. 1989. Thinking Through Talk: An Ethnographic Study of a Molecular Biology Laboratory. *Knowledge and Society* 8: 3–26.

Ackerman, S., and M.J. Hilsenroth. 2003. A Review of Therapist Characteristics and Techniques Positively Impacting the Therapeutic Alliance. *Clinical Psychology Review* 23 (1): 1–33.

Adams, J.L. 1979. *Conceptual Blockbusting*. 2nd ed. New York: Norton.

Angier, N. 1993. An Old Idea About Genius Wins New Scientific Support. *New York Times*, October 12. https://www.nytimes.com/1993/10/12/science/an-old-idea-about-genius-wins-new-scientific-support.html.

Armstrong, K. 2008. Charter for Compassion. https://charterforcompassion.org.

Augustine 1960. *The Confessions*. New York: Doubleday Image edition, pp. 397–400.

Aurobindo, S. 1990. *Synthesis of Yoga*. Twin Lakes, WI: Lotus Light Publications.

Bagnoli, M., H.A. Klein, C.G. Mann, and J. Robinson, eds. 2010. *Treasures of Heaven: Saints, Relics, and Devotion in Medieval Europe*. New Haven: Yale University Press.

Ball, P. 2011. Did Einstein Discover $E=mc^2$? (/$E=mc^2$/HistoryOfE=mC2A.webarchive).

Bakker, C., R. Gardner Jr., V. Koliatses, et al. 2002. The Social Brain: A Unifying Foundation for Psychiatry. *Letters, Academic Psychiatry* 6 (3): 219.

S. Restivo, *Einstein's Brain*,
https://doi.org/10.1007/978-3-030-32918-1

Balter, M. 2012. Rare Photos Show that Einstein's Brain Has Unusual Features. *The Washington Post*, Tuesday, 27 November E6.

Bargh, J.A., and T.L. Chartrand. 1999. The Unbearable Automaticity of Being. *American Psychologist* 54 (7): 462–479.

Barlow, A. 2013. *The Cult of Individualism: A History of an Enduring American Myth*. Santa Barbara, CA: Praeger.

Barnes, J., ed. 1984. *The Complete Works of Aristotle*. Vol. II. Princeton: Princeton University Press.

Baron-Cohen, S. 1997. *Mindblindness*. Cambridge, MA: MIT Press.

Barta, R. 2014. *Anthropology of the Brain: Consciousness, Culture, and Free Will*. Cambridge: Cambridge University Press.

Barthes, R. 2012. Einstein's Brain. In *Mythologies*, ed. R. Barthes, 2nd ed., 100–102. New York: Hill & Wang; orig. publ. in French, 1957.

Bartocci, U. 1999. *Albert Einstein & Olinto De Pretto*. Bologna: Societa Editrice, Andromeda.

Basalla, G. 1988. *The Evolution of Technology*. Cambridge: Cambridge University Press.

Battersby, C. 1989. *Gender and Genius: Towards a Feminist Aesthetics*. Bloomington: Indiana University Press.

Bealer, G. 1998. Intuition and the Autonomy of Philosophy. In *Rethinking Intuition: The Psychology of Intuition and Its Role in Philosophical Inquiry*, ed. M. Depaul and W. Ramsey, 201–239. Lanham, MD: Rowman & Littlefield.

Becker, G. 2000. The Association of Creativity and Psychopathology: Its Cultural-Historical Origins. *Creativity Research Journal* 13 (1): 45–53.

———. 2014. A Socio-Historical Overview of the Creativity-Pathology Connection from Antiquity to Contemporary Times. In *Creativity and Mental Illness*, ed. J.C. Kaufman, 3–24. Cambridge: Cambridge University Press.

Beretta, M., M. Confori, and P. Mazzarello, eds. 2016. *Savant Relics: Brains and Remains of Scientists*. Sagamore Beach, MA: Science History Publications.

Bertolero, M., and D. Bassett 2019. How Matter Becomes Mind. *Scientific American* July: 26–33.

Bodanis, D. 2009. *E = mc²: A Biography of the World's Most Famous Equation*. New York: Bloomsbury.

Bohm, D. 1971. *Causality and Chance in Modern Physics*. Philadelphia: University of Pennsylvania Press.

Bracken, P., P. Thomas, S. Timimi, et al. 2012. Psychiatry Beyond the Current Paradigm. *The British Journal of Psychiatry* 201: 430–434.

Brody, A.L., et al. 2001. Regional Brain Metabolic Changes in Patients With Major Depression Treated With Either Paroxetine or Interpersonal Therapy: Preliminary Findings. *Archives of General Psychiatry* 58 (7): 631–640.

Brothers, L. 1990. The Social Brain: A Project for Integrating Primate Behavior and Neurophysiology in a New Domain. *Concepts in Neuroscience* I: 27–51.

———. 1997. *Friday's Footprint: How Society Shapes the Human Mind*. New York: Oxford University Press.
———. 2001. *Mistaken Identity: The Mind-Brain Problem Reconsidered*. Albany, NY: SUNY Press.
Browne, M.W. 1987. Mathematics and Magic. *The New York Times Magazine*, October 18 (nytimes.com).
Brüne, M., H. Ribbert, and W. Schiefenhövel, eds. 2003. *The Social Brain: Evolution and Pathology*. West Sussex: John Wiley & Sons.
Burns, J. 2006. The Social Brain Hypothesis of Schizophrenia. *World Psychiatry* 5 (2): 77–81.
Cacioppo, J.T., G.G. Berntson, R. Adolphs, et al., eds. 2002. *Foundations in Social Neuroscience*. Cambridge, MA: MIT Press.
Callero, P. 2013. *The Myth of Individualism: How Social Forces Shape Our Lives*. Lanham, MD: Rowman & Littlefield.
Campbell, D.T. 1960. Blind Variation and Selective Retention in Creative Thought as in Other Knowledge Processes. *Psychological Review* 67: 380–400.
Carey, B. 2010. Genes as Mirrors of Life Experiences. *The New York Times*, November 8. https://www.nytims.com/2010/11/09/health/09brain.html.
Carhart-Harris, R.L., and J. Friston. 2019. REBUS and the Anarchic Brain: Toward a Unified Model of the Brain Action of Psychedelics. *Pharmacological Reviews* 71 (3): 316–344.
Caruso, G.D. 2012. *Free Will and Consciousness*. Lanham, MD: Lexington Books.
Charlton, B.G. 2003. Theory of Mind Delusions and Bizarre Delusions in an Evolutionary Perspective: Psychiatry and the Social Brain. In *The Social Brain: Evolution and Pathology*, ed. M. Brüne, H. Ribbert, and W. Schiefenhövel, 315–338. New York: John Wiley & Sons.
Chopra, D., and J. Orloff. 2001. *The Power of Intuition* (Audio CD). New York: Hay House.
Clark, A. 1998. *Being There: Putting Brain, Body, and World Together Again*. Cambridge, MA: Bradford.
Clark, R. 1971/1984. *Einstein: The Life and Times*. New York: Avon Books.
Clarke, G., et al. 2013. The Microbiome-Gut-Brain Axis During Early Life Regulates the Hippocampal Serotonergic System in a Sex-Dependent Manner. *Molecular Psychiatry* 18 (6): 666–673.
Clarke, J.M. 2015. *Where the Change Is: Everyday Interaction Rituals of Therapeutic Communities*. Doctoral Dissertation. Nottingham: University of Nottingham.
———. 2017a. The Case for "Fluid" Hierarchies in Therapeutic Communities. *Therapeutic Communities: The International Journal of Therapeutic Communities* 38 (4): 207–216.
———. 2017b. The Role of Everyday Interaction Rituals within Therapeutic Communities. In *Mental Health: Uncertainty and Inevitability: Rejuvenating the Relationship Between Social Science and Psychiatry*, ed. H. Middleton and M. Jordan, 47–72. London: Palgrave Macmillan.

Clarke, J.M., J. Waring, and S. Timmons. 2019. The Challenge of Inclusive Coproduction: The Importance of Situated Rituals and Emotional inclusivity in the. Coproduction of Health Research Projects. *Social Policy and Administration* 53 (2): 233–248.

Clarke, J.M., and J. Waring. 2018. The Transformative Role of Interaction Rituals Within Therapeutic Communities. *Sociology of Health and Illness* 40 (8): 1277–1293.

Clarke, S.P., J.M. Clarke, R. Brown, and H. Middleton. 2016. Hurting and Healing in Therapeutic Environments: How Can We Understand the Role of the Relational Context? *European Journal of Psychotherapy & Counselling* 18 (4): 384–400.

Collins, H., and T. Pinch. 1998. *The Golem: What You Should Know About Science.* 2nd ed. Cambridge: University of Cambridge Press; orig. publ. 1993.

Collins, R. 1992. *Sociological Insight: An Introduction to Non-Obvious Sociology.* 2nd ed. Oxford: Oxford University Press.

———. 1998. *The Sociology of Philosophies.* Cambridge, MA: Harvard University Press.

———. 2004. *Interaction Ritual Chains.* Princeton: Princeton University Press.

———. 2014. Interaction Ritual Chains and Collective Effervescence. In *Collective Emotions*, ed. C. von Scheve and M. Salmela, 299–311. Oxford: Oxford University Press.

Collins, R., and S. Restivo. 1983. Robber Barons and Politicians in Mathematics: A Conflict Model of Science. *The Canadian Journal of Sociology* 8 (2): 199–227.

Colombo, J., et al. 2006. Cerebral Cortex Astroglia and the Brain of a Genius: A Propos of A. Einstein's. *Brain Research Reviews* 52 (2): 257–263.

Connolly, J.C. 1950. *External Morphology of the Primate Brain.* Springfield, IL: C. C. Thomas.

Corry, L., Jürgen Renn, and John Stachel. 1997. Belated Decision in the Hilbert-Einstein Priority Dispute. *Science* 278 (5341): 1270–1273.

Cox, B., and J. Forshaw. 2009. *Why Does $E=mc^2$?* Boston: Da Capo Press.

Crane, M.T. 2000. *Shakespeare's Brain.* Princeton: Princeton University Press.

Critchlow, H. 2019. *The Science of Fate: Why Your Future Is More Predictable Than You Think.* Hachette, UK: Hodder & Stoughton.

Cruz, J.C. 1984. *Relics.* Huntington, IN: Our Sunday Visitor.

Csikaentmihaly, M. 1996. *Creativity: Flow and the Psychology of Discovery and Invention.* New York: Harper Perennial.

Damasio, A. 1994. *Descartes' Error.* New York: G.P. Putnam.

Davis, E. 1989. *Women's Intuition.* Berkeley, CA: Celestial Arts.

Dean, S. 2018. The Human Brain Can Create Structures in Up to 11 Dimensions. https://www.sciencealert.com/science-discovers-human-brain-works-up-to-11-dimensions.

DeNora, T. 1993. *Beethoven and the Construction of Genius: Musical Politics in Vienna, 1792–1803.* Berkeley: University of California Press.

DePaul, M., and W. Ramsey. 1998. *Rethinking Intuition: The. Psychology of Intuition and its Role in Philosophical Inquiry.* Lanham, MD: Rowman & Littlefield.

Diamond, M.C., A.B. Scheibel, G.M. Murphy, and T. Harvey. 1985. On the Brain of a Scientist: Albert Einstein. *Experimental Neurology* 88: 198–204.

Doyle, N. 2018. Gertrude Stein and the Domestication of Genius in The Autobiography of Alice B. Toklas. *Feminist Studies* 44 (1): 43–69.

Duke, D.W., and W.S. Pritchard, eds. 1991. *Measuring Chaos in the Brain.* London: World Scientific.

Dumit, J. 2004. *Picturing Personhood: Brain Scans and Biomedical Identity.* Princeton: Princeton University Press.

Durkheim, E. 1995. *The Elementary Forms of Religious Life.* Trans. K.E. Fields. New York: The Free Press, orig. publ. in French, 1912.

Dunbar, R. 1998. The Social Brain Hypothesis. *Evolutionary Anthropology* 1 (8): 184–190.

Dunbar, R., and S. Shultz. 2007. Evolution in the Social Brain. *Science* 317: 1344–1351.

Dunbar, R., C. Gamble, and J.G. Owlett, eds. 2010. *Social Brain: Distributed Mind.* Oxford: Oxford University Press.

Edmonds, M. 2008. How Albert Einstein's Brain Worked. *HowStuffWorks.com*, October 27. https://science.howstuffworks.com/life/inside-the-mind/human-brain/einsteins-brain.htm.

Elfenbein, A. 1996. Lesbianism and Romantic Genius: The Poetry of Anne Bannerman. *English Literary History* 63 (4): 929–957.

Falk, D., F.E. Lepore, and A. Noe. 2013. The Cerebral Cortex of Albert Einstein: A Description and Preliminary Analysis of Unpublished Photographs. *Brain: A Journal of Neurology* 136 (4): 1304–1327.

Fallon, J. 2013. *The Psychopath Inside: A Neuroscientist's Personal Journey into the Dark Side of the Brain.* New York: Current/Penguin.

Ferrari, P.F., and G. Rizzolatti. 2014. Mirror Neuron Research: The Past and the Future. *Philosophical Transactions of the Royal Society B* 369 (1644). https://doi.org/10.1098/rstb.2013.0169.

———. 2015. *New Frontiers in Mirror Neurons Research.* Oxford: Oxford University Press.

Feynman, R. 1979. *The Smartest Man in the World.* OMNI interview; reprinted as pp. 189–204 in Richard Feynman (1999). *The Pleasure of Finding Things Out.* New York: Basic Books.

Fiske and Taylor. 2013. *Social Cognition from Brains to Culture.* London: Sage.

Fleras, A. 2006. Ishi in Two Worlds: A Biography of the Last Wild Indian in North America. *Journal of Multilingual and Multicultural Development* 27 (3): 265–268.

Franklin, S. 1995. *Artificial Life.* Cambridge, MA: MIT Press.

Franks, D.D. and T. Smith, eds. (1999), Mind, Brain, and Society: Towards a Neurosociology of Emotion, Vol. 5 of Social Perspectives on Emotion (Stamford, CT: JAI Press).

Franks, D.D., and J.H. Turner, eds. 2013. *Handbook of Neurosociology*. New York: Springer.

French, P.J. 2002. *John Dee: The World of the Elizabethan Magus*. New York: Routledge; reprint of the 1972 Ark Paperback.

Galaburda, A. 1999. Albert Einstein's Brain. *Lancet* 354 (9192): 1821–1823.

Gardner, S. 2009. Nietzsche, the Self, and the Disunity of Philosophical Reason. In *Nietzsche on Freedom and Autonomy*, ed. Ken Gemes and Simon May, 1–31. Oxford: Oxford University Press.

Golden, F. 1999. Albert Einstein: Person of the Century. *Time* 154 (27): 62–65.

Gorney, R. 1972. *The Human Agenda*. New York: Simon & Schuster.

Guicciardini, N. 1999. *Reading the Principia: The Debate on Newton's Mathematical Methods for Natural Philosophy from 1687–1736*. Cambridge: Cambridge University Press.

Garber, Megan 2018. David Foster Wallace and the Dangerous Romance of Male Genius on the "Centrifugal Forces of Talented Men". *The Atlantic*, May 9. https://www.theatlantic.com/entertainment/archive/2018/05/the-world-still-spins-around-male-genius/559925/.

Garber, Marjorie. 2000. Our Genius Problem. *The Atlantic*, December, 2002: 65–72. https://www.theatlantic.com/magazine/archive/2002/12/our-genius-problem/308435/.

Gaver, W.W. 1991. Technology Affordances. In *Proceedings of the CHI*, 79–84. New York: ACM Press.

———. 1996. Affordances for Interaction: The Social Is Material for Design. *Ecological Psychology* 8 (2): 111–129.

Geertz, C. 1973. *The Interpretation of Cultures*. New York: Basic Books.

———. 2000. *Available Light*. Princeton: Princeton University Press.

Giannini, A.J., J. Daood, et al. 1978. Intellect versus Intuition-Dichotomy in the Reception of Nonverbal Communication. *Journal of General Psychology* 99: 19–24.

Gibson, J.J. 1966. *The Senses Considered as Perceptual Systems*. London: Allen & Unwin.

Gleick, J. 1992. *Genius: The Life and Science of Richard Feynman*. New York: Vintage.

Gumplowicz, L. 1980. *The Outlines of Sociology*. New Brunswick, NJ: Transaction Books, 1980; orig. publ. in German, 1885.

Hagmann, P. 2005. *From Diffusion MRI to Brain Connectomics Hampshire*. PhD Thesis. Lausanne: Ecole Polytechnique Fédérale de Lausanne.

Haight, R. 2013. The Quintessence of a Therapeutic Environment. *The International Journal of Therapeutic Communities* 34 (1): 6–15.

Hallett, T. 2003. Emotional Feedback and Amplification in Social Interaction. *The Sociological Quarterly* 44 (4): 705–726.

Halperin, J.M., and D.M. Healey. 2011. The Influences of Environmental Enrichment, Cognitive Enhancement, and Physical Exercise on Brain Development: Can we Alter the Developmental Trajectory of ADHD? *Neuroscience & Biobehavioral Reviews* 35: 621–634.

Hampshire, A., R.R. Highfield, et al. 2012. Fractionating Human Intelligence. *Neuron* 76 (6): 1225–1237.

Herculano-Houzel, S. 2014. The Glia/Neuron Ratio: How It Varies Uniformly Across Brain Structures and Species and What That Means for Brain Physiology and Evolution. Wiley Online Library (wileyonlinelibrary.com). https://doi.org/10.1002/GLIA.22683.

Hessen, B. 1931. The Social and Economic Roots of Newton's Principia. In *Science at the Crossroads*, ed. N.I. Bukharin et al., 151–212. London: Frank Cass and Company.

Highfield, B.L. Parkin, and A.M. Owen. 2012. Fractionating Human Intelligence. *Neuron* 76 (6): 1225–1237.

Hines, T.H. 2014. Neuromythology of Einstein's Brain. *Brain Cognition* 88 (July): 21–25.

Ho, P., and D.A. Ross. 2017. More Than a Gut Feeling: The Implications of the Gut Microbiota in Psychiatry. *Biological Psychiatry* 81 (5): e35–e37.

Holmes, E.A., M.G. Craske, and A.M. Graybiel. 2014. A Call for Mental Health Science. *Nature* 511: 287–289.

Hone, M. 2015. Gay Genius: From Plato to Nietzsche to Byron (© Michael Hone).

Hotz, R.L. 2015. Revealing Thoughts on Gender and Brains. *Los Angeles Times*, July 3.

Hofstadter, R. 1944. *Social Darwinism in American Thought*. Boston: Beacon Press.

Howe, M.J.A. 1989. *Fragments of Genius: The Strange Feats of Idiot Savants*. London: Routledge.

———. 1999. *Genius Explained*. Cambridge: Cambridge University Press.

Hsu, C. 2012. Scientists Find Truth in "Mad Scientist" Stereotype: There Is a Link between Genius and Insanity. Medicaldaily.com, June 4.

Hyeonlin, J., and L. Seung-Hwan. 2018. From Neurons to Social Beings: Short Review of the Mirror Neuron System Research and its Socio-Psychological and Psychiatric Implications. *Clinical Psychopharmacological Neuroscience* 16 (1): 18–31.

Isaacson, W. 2008. *Einstein: His Life and Universe*. New York: Simon & Schuster.

Jacob, F. 1977. Evolution and Tinkering. *Science* 196 (4295): 1161–1166.

Jacobs, B., and A.B. Scheibel. 1993. A Quantitative Dendritic Analysis of Wernicke's Area in Humans. I. Lifespan Changes. *Journal of Comparative Neurology* 327: 83–96.

Jacobs, B., M. Schall, and A.B. Scheibel. 1993. A Quantitative Dendritic Analysis of Wernicke's Area in Humans. II. Hemispheric and Environmental Factors. *The Journal of Comparative Neurology* 327: 97–111.

Jamison, K.R. 1989. Mood Disorders and Patterns of Creativity in British Writers and Artists. *Psychiatry* 52 (2): 125–134.

———. 1993. *Touched With Fire: Manic Depressive Illness and the Artistic Temperament*. New York: The Free Press.

———. 1995. *An Unquiet Mind*. New York: Alfred A. Knopf.

———. 2017. *Robert Lowell: Setting the River on Fire*. New York: Alfred A. Knopf.

Janik, A., and S.E. Toulmin. 1973. *Wittgenstein's Vienna*. New York: Simon & Schuster.

Jenkins, T. 2013. De (Re) Constructing Ideas of Genius: Hip Hop, Knowledge, and Intelligence. *International Journal of Critical Pedagogy* 4 (3): 11–23.

Joffe, J.M., et al. 1984. *Readings in Primary Prevention of Psychopathology: Basic Concepts*. Lebanon, NH: University Press of New England).

Johnson-Ulrich, L. 2017. The Social Intelligence Hypothesis. In *Encyclopedia of Evolutionary Psychological Science*, ed. T.K. Shackelford and V.A. Weekes-Shackelford, 1–7. New York: Springer.

Johnston, H. 2009. Newton's Wars. *Everyday Science/Blog*, September. https://physicsworld.com/a/newtons-wars-ii/.

Jones, M. 1976. *Maturation of the Therapeutic Community: An Organic Approach to Health and Mental Health*. New York: Human Sciences Press.

Kalb, C. 2017. What Makes a Genius? *National Geographic* 231 (5): 30–55.

———. 2018. *The Science of Genius*. National Geographic: Special Publication.

Kahneman, D. 2011. *Thinking, Fast and Slow*. New York: Farrar, Strauss, and Giroux.

Kandel, E.R. 1998. A New Intellectual Framework for Psychiatry. *American Journal of Psychiatry* 155: 457–469.

Kandel, E.R., J.H. Schwartz, et al., eds. 2013. *Principles of Neural Science*. 5th ed. New York: McGraw-Hill.

Kandel, E.R. 2018. *The Disordered Mind: What Unusual Brains Tell Us About Ourselves*. New York: Farrar, Strauss, and Giroux.

Kantha, S.S. 1992. Albert Einstein's Dyslexia and the Significance of Brodmann Area 39 of His Left Cerebral Cortex. *Medical Hypotheses* 37 (2): 119–122.

Kaptelinin, V. 2014. *Affordances and Design*. Aarhus N, Denmark: The Interaction Design Foundation.

Keller, E.F. 1983. *A Feeling for the Organism: The Life and Work of Barbara McClintock*. New York: W.H. Freeman.

Kennedy, D.P., and R. Adolphs. 2012. The Social Brain in Psychiatric and Neurological Disorders. *Trends in Cognitive Sciences* 11: 559–572.

Kessler, M., and S.E. Goldston. 1986. *A Decade of Progress in Primary Prevention*. Lebanon, NH: University Press of New England.

Konnikova, M. 2014. How Headlines Change the Way We Think. *The New Yorker*. https://www.newyorker.com/science/maria-konnikova/headlines-change-way-think.

Kroeber, T. 1961. *Ishi in Two Worlds*. Berkeley: University of California Press.

Krubitzer, L. 2014. Lessons from Evolution. In *The Future of the Brain*, ed. G. Marcus and J. Freeman, 186–193. Princeton: Princeton University Press.

Kull, K. 1998. On Semiosis, Umwelt, and Semiosphere. *Semiotica* 120 (3/4): 299–310.

Kuhn, T.S. 1962. *The Structure of Scientific Revolutions*. Chicago: University of Chicago Press, 2nd ed., 1970.

Kyaga, S. 2015. *Creativity and Mental Illness: The Mad Genius in Question*. London: Palgrave Macmillan.

Khanna, P. 2016. *Connectography: Mapping the Future of Global Civilization*. New York: Random House.

Kinderman, P. 2014. *Prescription for Psychiatry: Why We Need a Whole New Approach to Mental Health and Wellbeing*. New York: Palgrave Macmillan.

King, B. 2007. *Evolving God*. New York: Doubleday.

Kivy, P. 2001. *The Possessor and the Possessed: Handel, Mozart, Beethoven and the Idea of Musical Genius*. New Haven: Yale University Press.

Klein, G.A. 1998. *Sources of Power: How People Make Decisions*. Cambridge, MA: MIT Press.

———. 2003. *Intuition at Work*. New York: Random House.

Koh, C. 2006. Reviewing the Link Between Creativity and Madness: A Postmodern Perspective. *Educational Research and Reviews* 1 (7): 213–221.

Koberlein, B. 2017. The History of Einstein's Most Famous Equation. https://briankoberlein.com.

Kroeber, A. 1963. *Configurations of Culture Growth*. Berkeley: University of California Press.

Langer, E.J. 1978. Rethinking the Role of Thought in Social Interaction. In *New directions in attribution research*, ed. J.H. Harvey, W. Ickes, and R.F. Kidd, vol. 2, 35–58. Hillsdale, NJ: Lawrence Erlbaum Associates.

Langer, E., A. Blank, and B. Chanowitz. 1978. The Mindlessness of Ostensibly Thoughtful Action: The Role of "Placebic" Information in Interpersonal Interaction. *Journal of Personality and Social Psychology* 36: 635–642.

Larsen, R.D. 1993. Kekulé's Benzolfest Speech: A Fertile Resource for the Sociology of Science. In *The Kekulé Riddle: A Challenge for Chemists and Psychologists*, ed. J.H. Wotiz, 177–193. Vienna, IL: Cache River Press.

Lehnertz, K., C.E. Elger, J. Arnhold, and P. Grassberger, eds. 2000. *Chaos in Brain? Proceedings of the Workshop*. University of Bonn: Germany.

Lepore, F.E. 2018. *Finding Einstein's Brain*. New Brunswick, NJ: Rutgers University Press.

Levy, S. 1978. I Found Einstein's Brain. *New Jersey Monthly*, August. https://njmonthly.com/articles/historic-jersey/the-search-for-einsteins-brain/.

Libet, B., C.A. Gleason, E.W. Wright Gleason, and D.K. Pearl. 1983. Time of Conscious Intention to Act in Relation to Onset of Cerebral Activity (Readiness-Potential) – The Unconscious Initiation of a Freely Voluntary Act. *Brain.* 106 (3): 623–642.

Libet, B. 1985. Unconscious Cerebral Initiative and the Role of Conscious Will in Voluntary Action. *The Behavioral and Brain Sciences.* 8 (4): 529–566.

Libet, B., A. Freeman, and K. Sutherland. 1999. *The Volitional Brain: Toward a Neuroscience Free Will.* Exeter: Imprint Academic.

Libet, B. 2004. *Mind Time: The Temporal Factor in Consciousness.* Cambridge, MA: Harvard University Press.

Lombroso, C. 1891. *The Man of Genius.* London: Walter Scott.

Long, P., and G. Corfas. 2014. Dynamic Regulation of Myelination in Health and Disease. *JAMA Psychiatry* 71 (11): 1296–1297.

Lynch, M. 2006. Trusting Intuitions. In *Truth and Realism*, ed. P. Greenough and M. Lynch, 227–238. Oxford: Oxford University Press.

Martin, P., ed. 1969. *Genius: The History of an Idea.* New York: Basil Blackwell.

Maturana, H.R., and F. Varela. 1980. *Autopoiesis and Cognition.* Dordrecht: D. Reidel.

Marano, H.F. 2007. Creativity and Mood: The Myth that Madness Heightens Creative Genius. https://www.psychologytoday.com/us/articles/200705/genius-and-madness?quicktabs_5=1.

Markram, H. 2008. Fixing the Location and Dimensions of Functional Neocortical Columns. *Human Frontiers Science Program Journal* 2 (3): 132–135.

McAlpine, D.D., and D. Mechanic. 2011. The Influence of Social Factors on Mental Health. In *Principles and Practice of Geriatric Psychiatry*, ed. M.T. Abo-Saleh, C.L.E. Katona, and A. Kumar, 3rd ed., 97–102. New York: John Wiley & Sons.

McMahon, D.M. 2013. *Divine Fury: A History of Genius.* New York: Basic Books.

Mead, G.H. 1934. *Mind, Self, & Society.* Vol. 1. Chicago: University of Chicago Press.

Men, W., D. Falk, T. Sun, W. Chen, J. Li, D. Yin, L. Zang, and M. Fan. 2013. The Corpus Callosum of Albert Einstein's Brain: Another Clue to his High Intelligence? *Brain: A Journal of Neurology* 137 (4): e268. https://doi.org/10.1093/brain/awt252.

Mercier, H., and D. Sperber. 2017. *The Enigma of Reason.* Cambridge, MA: Harvard University Press.

Merton, R.K. 1961. Singletons and Multiples in Scientific Discovery: A Chapter in the Sociology of Science. *Proceedings of the American Philosophical Society*, 105: 470–486; reprinted in R.K. Merton (1973). *The Sociology of Science.* Chicago: University of Chicago Press, 1973, pp. 343–370.

———. 1965. *Standing on the Shoulders of Giants: A Shandean Postscript.* New York: The Free Press.

———. 1968. *Social Theory and Social Structure*. New York: The Free Press.
———. 1968a. On the History and Systematics of Sociological Theory. Chapter 1 in *Social Theory and Social Structure*, enlarged ed. New York: The Free Press.
———. 1968b. Science and Economy of 17th Century England. In *Social Theory and Social Structure*, ed, R.K. Merton, enlarged ed., 661–681. New York: The Free Press; orig. publ. 1939.
———. 1973. *The Sociology of Science*. Chicago: University of Chicago Press.
Mialet, H. 2010. *Hawking Incorporated: Stephen Hawking and the Anthropology of the Knowing Subject*. Chicago: University of Chicago Press.
Miller, A.I. 2001. *Einstein, Picasso: Space, Time, and the Beauty That Causes Havoc*. New York: Basic Books.
Monk, R. 1990. *Ludwig Wittgenstein: The Duty of Genius*. New York: Penguin.
Montagu, A. 1952. *Darwin: Competition and Cooperation*. New York: Henry Schuman.
Moore, D. 2002. *The Dependent Gene: The Fallacy of "Nature vs. Nurture"*. New York: Henry Holt and Company.
———. 2015. *The Developing Genome: An Introduction to Behavioral Epigenetics*. Oxford: Oxford University Press.
Neihart, M. 1998. Creativity, the Arts, and Madness. *Roeper Review* 21: 47–50.
Nietzsche, F. 1996/1878–1880. *Human, All Too Human*. Cambridge: Cambridge University Press.
Noé, A. 2010. *Out of Our Heads: Why You Are Not Your Brain, and Other Lessons from the Biology of Consciousness*. New York: Hill & Wang.
Norman, D. 1988. *The Psychology of Everyday Things*. New York: Basic Books.
Norretranders, T. 1991. *The User Illusion: Cutting Consciousness Down to Size*. New York: Viking.
Okasha, A. 2009. The Social Brain: A New Perspective. *Current Psychiatry* 16 (1): 1–6.
Orlinsky, D.E., and K.J. Howard. 1986. Process and Outcome in Psychotherapy. In *Handbook of Psychotherapy and Behavior Change*, ed. S. Garfield and A. Bergin, 3rd ed., 311–381. New York: John Wiley & Sons.
Pang, T., and A. Hannan. 2012. Enhancement of Cognitive Function in Models of Brain Disease Through Environmental Enrichment and Physical Activity. *Neuropharmacology* 64: 515–528.
Parayil, G. 1999. *Conceptualizing Technological Change*. Lanham, MD: Rowman & Littlefield.
Paterniti, M. 2000. *Driving Mr. Albert: A Trip Across America with Einstein's Brain*. New York: Dial Press.
Pelphrey, K.A., C.M. Hudac, et al. 2011. Research Review: Constraining Heterogeneity: The Social Brain and its Development in Autism Spectrum Disorder. *Journal of Child Psychology and Psychiatry* 52 (60): 631–644.

Perelman, B. 1994. *The Trouble With Genius: Reading Pound, Joyce, Stein, and Zukovsky*. Berkeley: University of California Press.
Pescosolido, B. 2013. Organizing the Sociological Landscape for the Next Decades of Health and Health Care Research: The Network Episode Model III-R As Cartographic Subfield Guide. In *Handbook of the Sociology of Health, Illness, and Healing*, ed. B. Pescosolido, J. Martin, J.D. McLeod, and A. Rogers, 39–66. New York: Springer.
Pickersgill, M., and I.V. Keulen, eds. 2012. *Sociological Reflections on the Neurosciences*, Advances in Medical Sociology. Vol. 13. Bingley, UK: Emerald Group Publishing, Ltd.
Poe, E.A. 1903/1842. Eleonora. Opening lines in *The Works of Edgar Allan Poe*, The Raven Edition, 5 Vols. New York: P.F. Collier and Son.
Preston, S.T. 1875. *Physics of the Ether*. London: E. & F.N. Spon.
Putnam, H. 1981. *Reason, Truth, and History*. Cambridge: Cambridge University Press.
Racine, V. 2014. Evolution and tinkering (1977), by Francois Jacob. Embryo Project Encyclopedia. https://embryo.asu.edu/pages/evolution-and-tinkering-1977-francois-jacob.
Radhakrishnan, S. 1932. *Idealist View of Life*. New York: Simon & Schuster.
Rathore, H. 2018. Why Did Einstein Use Speed of Light Squared in the Famous Equation $E = mc^2$? https://www.quora.com/Why-did-Einstein-use-speed-of-light-squared-in-the-famous-equation-E-mc-2.
Reimann, M.W., M. Nolte, et al. 2017. Cliques of Neurons Bound into Cavities Provide a Missing Link Between Structure and Function. *Frontiers in Computational Neuroscience* 11 (48): 1–16.
Restivo, S. 1983. The Myth of the Kuhnian Revolution in the Sociology of Science. In *Sociological Theory*, ed. R. Collins, 293–305. New York: Jossey-Bass.
———. 1992. *Mathematics in Society and History*. New York: Springer.
———. 1994. *Science, Society, and Values: Toward a Sociology of Objectivity*. Bethlehem, PA: Lehigh University Press.
———. 2005/2019. Romancing the Robots: Social Robots and Society, pp. 10, 17. Available at salrestivo.org.
———. 2008. Society, Social Construction, and the Sociological Imagination. *Constructivist Foundations* 3 (2): 94–96.
———. 2016. *Red, Black and Objective*. New York: Routledge.
———. 2017. *Sociology, Science, and the End of Philosophy: How Society Shapes Brains, Gods, Maths, and Logics*. New York: Palgrave Macmillan.
———. 2018. *The Age of the Social: The Discovery of Society and the Ascendance of a New Episteme*. New York: Routledge.
Restivo, S., and J. Croissant. 2007. Social Constructionism in Science and Technology Studies. In *Handbook of Constructionist Research*, ed. J. Holstein and J. Gubrium, 213–229. New York: Guilford Press.

Restivo, S., S. Weiss, and A. Stingl. 2014. *Worlds of ScienceCraft*. New York: Routledge.

Restivo, S., and S. Weiss. 2016. The Social Ecology of Brain and Mind. In *Worlds of ScienceCraft*, ed. S. Restivo, S. Weiss, and A. Stingl, 37–70. New York: Routledge.

Robson, M., and P. Miller. 2006. Australian Elite Leaders and Intuition. *Australasian Journal of Business and Social Inquiry* 4 (3): 43–61.

Root-Berstein, R. 1989. *Discovering, Inventing and Solving Problems at the Frontiers of Science*. Cambridge, MA: Harvard University Press.

Rose, S., and H. Rose. 2016. *Can Neuroscience Change Our Minds?* Cambridge: Polity Press.

Rossner, M. 2013. *Just Emotions: Rituals of Restorative Justice*. Oxford: Oxford University Press).

Roth, A., and P. Fongay. 2005. *What Works For Whom? A Critical Review of Psychotherapy Research*. New York: The Guilford Press.

Rothenberg, A. 1990. *Creativity & Madness*. Baltimore, MD: Johns Hopkins University Press.

———. 2014. *Flight from Wonder: An Investigation of Scientific Creativity*. New York: Oxford University. Press.

Rothman, T. 2015. Was Einstein the First to Invent E = mc²?(/E=mc2/Was%20Einstein%20the%20First%20to%20Invent%20E%20=%20mc2%3F%20-%20Scientific%20American.webarchive).

Rowan, J., and M. Cooper. 1999. *The Plural Self: Multiplicity in Everyday Life*. London: Sage.

Saks, E. 2007. *The Center Cannot Hold: My Journey Through Madness*. New York: Hyperion.

Schutt, R.K., L. Seidman, and M.S. Keshavan, eds. 2015. *Social Neuroscience: Brain, Mind, and Society*. Cambridge, MA: Harvard University Press.

Scott, K.M., A.O. Al-Hamzawi, L.H. Andrade, et al. 2014. Associations Between Subjective Social Status and DSM-IV Mental Disorders: Results from the World Mental Health Surveys. *JAMA Psychiatry* 71 (12): 1400–1408.

Sebeok, Thomas A. 1976. Foreword. In *Contributions to the Doctrine of Signs*, ed. T. Sebeok, x. Lanham, NH: University Press of America.

Sebeok, Thomas A., and D.J. Umlker-Sebeok. 1978. Linguistics: Speech Surrogates: Drum and Whistle Systems. *American Anthropologist* 80 (3): 712–713.

Schlesinger, J. 2009. Creative Myth Conceptions: A Closer Look at the Evidence for the 'Mad Genius' Hypothesis. *Psychology of Aesthetics, Creativity, and the Arts* 3 (2): 62–72.

Searle, J. 1992. *The Rediscovery of Mind*. Cambridge, MA: MIT Press.

Seung, S. 2012. *Connectome: How the Brain's Wiring Makes Us Who We Are*. New York: Houghton Mifflin Harcourt.

Shakespeare, W. 1998/1600. A Midsummer Night's Dream (Act V, Scene 1, openinglines).https://www.gutenberg.org/files/1514/1514-h/1514-h.htm.
Shapin, S. 1981. Licking Leibniz. *History of Science* 19: 293–305.
Shea, E.P. 2008. *How the Gene Got Its Groove*. Albany, NY: SUNY Press.
Shonkoff, J.P., and D.A. Phillips, eds. 2000. *From Neurons to Neighborhoods: The Science of Early Childhood Development*. Washington, DC: The National Academies Press.
Simonton, D.K. 1999. *Origins of Genius*. New York: Oxford University Press.
Skarda, C.A., and W.J. Freeman. 1987. How Brains Make Chaos in Order to Make Sense of the World. *Behavioral and Brain Sciences* 10: 161–195.
———. 1990. Chaos and the New Science of the Brain. *Concepts in Neuroscience* 2: 275–285.
Soegaard, M. 2015. Affordances. In *The Glossary of Human Computer Interaction*, ed. B. Papantoniou, M. Soegaard, J. Lupton, et al. Aarhus N, Denmark: The Interaction Design Foundation.
Sonar, T. 2018. *The History of the Priority Dispute Between Newton and Leibniz: Mathematics in History and Culture*. Cham, Switzerland: Birkhäuser, orig. in German, 2016.
Soresi, E. 2014. *The Anarchic Brain*. Milano: Bookrepublic.
Sporns, O., G. Tononi, and R. Kötter. 2005. The Human Connectome: A Structural Description of the Human Brain. *PLoS Computational Biology* 1 (4): e42.
Stadler, G. 1999. Louisa May Alcott's Queer Geniuses. *American Literature* 71 (4): 657–677.
Starns, O. 1994. *Ishi's Brain*. New York: W.W. Norton.
Steinmetz, H., A. Herzog, et al. 1994. Discordant Brain-Surface Anatomy in Monozygotic Twins. *New England Journal of Medicine* 331: 952–953.
Sztompka, P. 1991. *Society in Action: The Theory of Social Becoming*. Chicago: University of Chicago Press.
Subotnik, R.F., and K.D. Arnold. 1994. *Beyond Terman*. Norwood, NJ: Ablex Publishers.
Summers-Effler, E. 2002. The Micro Potential for Social Change: Emotion, Consciousness, and Social Movement Formation. *Sociological Theory* 20 (1): 41–60.
———. 2004a. A Theory of the Self, Emotion, and Culture. Theory and Research on Human Emotions. In *Theory and Research on Human Emotions*, ed. J. Turner, vol. 21, 273–308. Bingley, West Yorkshire, UK: Emerald Group Publishing, Ltd.
———. 2004b. Defensive Strategies: The Formation and Social Implications of Patterned Self-destructive Behavior. *Advances in Group Processes* 21: 309–325.
Swaab, D.F. 2014. *We Are Our Brains: A Neurobiography of the Brain, From the Womb to Alzheimer's*. New York: Spiegel & Grau.

Taylor, J.M. 2016. Mirror Neurons After a Quarter Century: New Light, New Cracks. http://sitn.hms.harvard.edu/flash/2016/mirror-neurons-quarter-century-new-light-new-cracks/.

Terman, L. 1925. *Mental and Physical Traits of a Thousand Gifted Children*, Vol. 1 of 5 in Genetic Studies of Genius. Stanford: Stanford University Press.

Thompson, E., and D. Cosmelli. 2011. Brain in a Vat or Body in a World? Brainbound Versus Enactive Views of Experience. *Philosophical Topics* 39 (1): 163–180.

Thorndike, L. 1923–1958. *A History of Magic and Experimental Science*, 8 vols. New York: Macmillan & Columbia University Press.

Trevarthen, C. 2000. Musicality and the Intrinsic Motive Pulse: Evidence from Human Psychology and Infant Communication. *Musicae Scientiae* 3 (1): 155–215.

Trevarthen, C., and K.J. Aitken. 2001. Infant Intersubjectivity: Research, Theory and Clinical Applications. *Journal of Child Psychology and Psychiatry* 42 (1): 3–48.

Two-Stage Models for Free Will, Retrieved June 27, 2019, from Information Philosopher, Web site http://www.informationphilosopher.com/freedom/twostage_models.html.

Ulam, S.M. 1976. *Adventures of a Mathematician*. New York: Charles Scribner's Sons.

von Uexküll, Thure. 1987. The Sign Theory of Jakob von Uexküll. In *Classics of Semiotics*, ed. M. Krampen et al., 147–179. New York: Plenum.

———. 1957. A Stroll Through the Worlds of Animals and Men: A Picture Book of Invisible Worlds. In *Instinctive Behavior: The Development of a Modern Concept*, ed. Claire H. Schiller, 5–80. New York: International Universities Press.

von Uexküll, J. 2010/1934. *A Foray into the World of Animals and Humans With a Theory of Meaning*. Trans. J.D. O'Neil. Minneapolis: University of Minnesota Press.

de Waal, P.L. Tyack, et al., eds. 2003. *Animal Social Complexity: Intelligence, Culture and Individualized Societies*. Cambridge, MA: Harvard University Press.

Waring, J., S. Bishop, J. Clarke, et al. 2018. Healthcare Leadership with Political Astuteness (HeLPA): A Qualitative Study of How Service Leaders Understand and Mediate the Informal 'Power and Politics' of Major Health System Change. *BMC Health Services Research* 18: 918.

Wegner, D.M. 2002. *The Illusion of Conscious Will*. Cambridge, MA: Bradford Books.

Weiner, E. 2016. *The Geography of Genius: Lessons from the World's Most Creative Places*. New York: Simon & Schuster.

Weisberg, R. 1986. *Creativity: Genius and Other Myths*. New York: W.H. Freeman.

———. 1993. *Creativity: Beyond the Myth of Genius*. New York: W.H. Freeman.

Westfall, R.S. 1980. Newton's Marvelous Years of Discovery and Their Aftermath. *ISIS* 71: 101–121.

Whitehead, C., ed. 2008. *The Origin of Consciousness in the Social World*. Exeter: Imprint Academic.

Whiten, A. 2000: 185–196; discussion 196–201. *Social Complexity and Social Intelligence*. Basel, Switzerland: Novartis Foundation Symposium 233.

Whitman, Walt. 2007/1855. *Leaves of Grass*. Mineola, NY: Dover Thrift Editions.

Whittaker, E. 1989. *The History of the Theories of Aether & Electricity*, Vols. 1 & 2. Mineola, NY: Dover Publications; orig. publ. 1951–1953.

Wilkinson, L.R., and B. Gerontol. 2016. Financial Strain and Mental Health Among Older Adults During the Great Recession. *Psychological Sciences and Social Sciences* 71 (4): 745–754.

Williams, D.L. 2002. *The Mind in the Cave*. High Holborn, UK: Thames and Hudson.

Wilson, E.A. 1998. *Neural Geographies*. New York: Routledge.

Wilson, E.O. 2013. *The Social Conquest of Earth*. New York: Liveright.

Witelson, S.F., D.L. Kigar, and T. Harvey. 1999. The Exceptional Brain of Albert Einstein. *The Lancet* 353 (9170): 2149–2153.

Winterberg, F. 2006. Response to Cory-Renn-Stachel. *Zeitschrift für Naturforschung* 59a: 715–719.

Wolchover, N. 2012. Why Are Genius and Madness Connected. https://www.livescience.com/20713-genius-madness-connected.html.

Wotiz, John H., and Susanna Rudofsky. 1993. Herr Professor Doktor Kekulé: Why Dreams? In *The Kekulé Riddle: A Challenge for Chemists and Psychologists*, ed. J.H. Wotiz, 247–276. Vienna, IL: Cache River Press.

Wurtele, S.K. 1998. Victims of Child Mistreatment. In *Comprehensive Clinical Psychology*, ed. A.S. Bellack and M. Hersen, vol. 9, 341–358. New York: Pergamon.

Zaidel, D. 2001. Neuron Soma Size in the Left and Right Hippocampus of a Genius. [Conference Poster: http://cogprints.org.1927].

Zapporoli, L., M. Porta, and E. Paulesu. 2015. The Anarchic Brain in Action: The Contribution of Task-Based fMRI Studies to the Understanding of Gilles de la Tourette Syndrome. *Current Opinion in Neurology* 6: 604–611.

Zetterberg, J.P. 1980. The Mistaking of 'the Mathematicks' for Magic and Tudor and Stuart England. *Sixteenth Century Journal* 11: 83–97.

Index[1]

[1] Note: Page numbers followed by 'n' refer to notes.

S. Restivo, *Einstein's Brain*,
https://doi.org/10.1007/978-3-030-32918-1